D1202188

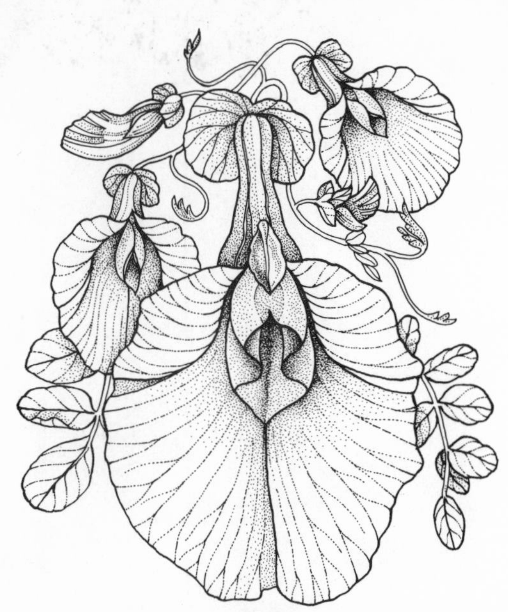

The Sex Life of Plants

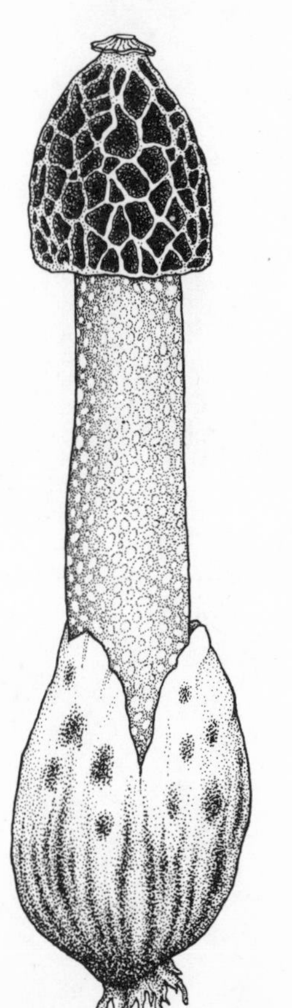

Frontispiece illustrations:
Above: *Clitoria ternatea*
Below: *Phallus impudicus*

The Sex Life of Plants

Alec Bristow

Holt, Rinehart and Winston
New York

To those who have
taught me about sex

Published simultaneously in Canada by
Holt, Rinehart and Winston of Canada, Limited.

Library of Congress Cataloging in Publication Data
Bristow, Alec.
The sex life of plants
Includes index.
1. Plants, Sex in. I. Title
QK827.B76 1978 581.3'6 77-8934
ISBN 0-03-022806-9

First Edition

Designers: Martin Bassett/Julian Holland
Illustrations: Paul Turner

Printed in the United States of America

10 9 8 7 6 5 4 3 2 1

Contents

Introduction

The plant kingdom invented sex. Long before the first simple forms of animal life began to appear in the steaming primeval oceans, the jungles had already, over millions of years, developed a fantastically rich variety of forms. From the giant tree ferns and cycads – forerunners of the trees – to the microscopic moss-like plants invisible to the naked eye (had there been eyes to see them in that dawn of time) they had acquired a vast amount of sexual experience. As we shall see in this book, the plants have tried, and are still trying, every conceivable form of sexual experiment. Some of their methods of sexual expression are indeed so peculiar that few forms of animal life would have either the imagination or the equipment to copy them.

But then plants have been at it for a very much longer time than animals. According to the Bible (Genesis 1, 11–13) the vegetable kingdom was made on the third day of creation, just after the seas had been divided from the dry land: "Let the earth bring forth grass, the herb yielding seed, and the fruit tree yielding fruit after his kind, whose seed is in itself, upon the earth: and it was so. And the earth brought forth grass and herb yielding seed after his kind, and the tree yielding fruit, whose seed was in itself, after his kind: and God saw that it was good. And the evening and the morning were the third day."

There are few people who would today believe those words as a literal statement of fact, but little more than a hundred years ago they were accepted without question by a great many intelligent people. In their *Flora,* published in 1860, two highly respected botanists, Hooker and Arnott, wrote "Many species were simultaneously called into existence on the third day of creation, each distinct from the other and destined to remain so." Yet the previous year, in 1859, Charles Darwin had published his *Origin of Species,* which was to overthrow the orthodox belief that six thousand years ago, in less than a week, God had created the world and every living thing in it, made up of a fixed number of kinds that could never be changed. It was largely as a result of his studies of plants and the many ways in which they satisfy their sexual urges that Darwin came to the conclusion that the number of species was not fixed but continually changing, new ones emerging and old ones becoming extinct according to their ability to adapt to circumstances.

It would now be generally agreed by scientists that sex has been the chief instrument of evolution; indeed, once sex had been invented evolution had become inevitable, natural selection being primarily based upon sexual selection.

Biologists are basically voyeurs, obtaining their satisfaction from watching the carryings-on of living creatures, especially their sexual carryings-on; and Darwin was one of the supreme voyeurs of all time, not only getting great personal satisfaction from what he saw but recording all the details with minute accuracy and obvious relish. His writings can still be read not only for interest but for pleasure.

However, for many years after the publication of Darwin's observations it still remained an article of faith among true believers that every species of plant had been simultaneously brought into being by God on the third day of creation and that no new ones had appeared since. Though members of some of the more extreme fundamentalist sects believe that still, in order to do so they must deliberately close their eyes to every fact of nature – expecially sexual nature. Nevertheless they continue to believe what almost the entire establishment asserted in Darwin's day: that the entire vegetable kingdom was dropped, intact and unchanging, upon the earth on the third day of creation. That was, according to the Bible story, the day before the creation of the sun, the moon and the stars. Not till the fifth day were the sea creatures and the birds made; and it was only on the sixth and final day that the land animals were created, including man, and the words at last appeared "male and female created he them."

For want of those words in relation to the making of plants, there were to be bitter theological denunciations of the seventeenth and eighteenth century botanists who first began to suggest that plants had a sex life. In vain the botanists, armed with such newfangled instruments as microscopes, peered into plants and described the sexual organs they saw there. Wicked lies, the biblical scholars said, and filthy wicked lies at that. The Bible was quite clear. It did *not* say of plants "male and female created he them." Without male and female there could be no sex. Therefore plants had no sex. And that was that. It was not the first time, nor was it to be the last, that scientists were attacked for drawing attention to facts that seemed to go against holy writ. Only forty years before the first suggestion about plant sex was made in a lecture to the learned members of the Royal Society in London, the great astronomer Galileo had been forced by the Inquisition in Rome, under threat of torture, to recant after his observation – made with the aid of another newfangled instrument called the telescope – that the earth goes around the sun. The consulting theologians of the Holy Office had advised the Pope that such an observation was opposed to biblical doctrine and therefore heretical.

By the time that the first suggestion about plant sex was made to the Royal Society, Galileo's "heresy" had very nearly been accepted as orthodox belief. The acceptance of plant sex has taken a good deal longer. To some extent that is because much remained to be discovered on the subject, and in fact still does. To an even greater extent it is because sex is a highly emotive subject which roused strong feelings of disapproval, fear and disgust among many members of the religious establishment. Astronomical and physical heresies could only be called heresies; sexual ones could be called obscene as well.

The first chapter of this book gives some account of the ignorance, prejudice and prudery which the discovery of plant sex has had to overcome before it could be accepted as a fact of life. The resistance to the very idea of sexuality in the vegetable kingdom has been threefold. First, new discoveries are in themselves unwelcome to the establishment, because they are bound to call existing beliefs into question or at least to require the exercise of critical faculties, and nothing more thoroughly blunts the critical faculties than blind faith. Second, as we have seen, no mention of sex is made in the Bible until the sixth day; so how, the faithful asked, could it apply to plants, which were made on the third day? Lastly, as we shall see, the whole notion of sex in relation to plants has been distasteful for psychological reasons. To those who only grudgingly accept the facts of sex, plants represent something pure and unspotted; the suggestion of sexual drives in the vegetable kingdom was as outrageous to such people as the later discoveries by Freud and other psychologists that infants had sexual drives as well, instead of being pure and innocent.

As Alex Comfort has pointed out in his illuminating book *Sex in Society*, disapproval of sex, sexual discoveries and books about sex is most vocally expressed by the most disturbed and prohibitive people in the community. They are, to quote the philosopher Ouspensky, the least sexually active section of society, the "infra-sex," who tend to compensate for their animal deficiencies by developing an abnormally strong desire to regulate the conduct of others.

All enlightenment tends to be resented and resisted by such people; after all, they will say, the Fall of man was caused by eating the fruit of the tree of knowledge, so the pursuit of knowledge is itself wrong and inspired by the Devil. However, Devil-inspired or not, the pursuit of knowledge goes on, and in the three centuries since plant sex was first discovered, a vast amount of information has been amassed on the subject. This book can only present a selection of some of the facts that have been established up to the present. A great deal remains to be discovered, and no doubt new findings in the future will in their turn call our present beliefs on the subject into question.

Two things remain clear. First, sex is a mark of the higher forms of life, whether plant or animal; it is the sexless ones that are primitive. Second, when plants invented sex they invented beauty. Every manifestation of beauty – whether of color, form, touch, scent or taste – is based upon sexual attraction, for which we are indebted originally to the plant world.

"From giant Oaks, that wave their branches dark,
To the dwarf Moss that clings upon their bark,
What Beaux and Beauties crowd the gaudy groves,
And woo and win their vegetable loves."

– Erasmus Darwin

The Loves of the Plants, from *The Botanic Garden, a poem in two parts with Philosophical Notes.*
First edition 1789.

I

The Discovery of Sex

In 1916 an article appeared in the *Journal of the National Horticultural Society of France* under the heading "A Curious Case of Mimicry among the Orchids." It described how a Frenchman in Algiers named Pouyanne had noticed a strange thing. During working hours M. Pouyanne was a Conseilleur – a sort of judge-advocate – at the Court of Appeal; during his spare time he was a keen amateur naturalist. The article told how he was watching the flowers of a species of orchid called *Ophrys speculum* and wondering, as many other people had wondered before, why the lip of each flower so closely resembled a certain kind of wasp which often could be seen visiting the orchids, and why it was only male wasps that visited them. He watched a wasp alight on one of the flowers, cling to it and perform some strenuous, jerky movements. The wasp had obviously become highly excited, and so by now had the judge.

What was going on? What satisfaction was the wasp – and for that matter the flower – getting from the encounter? At first sight it might have appeared that the wasp was attacking the flower for some obscure reason, but the judge was a careful observer and soon realized that the driving force behind the wasp's behavior was not violence but sex, not hate but love. He had suspected something of the sort before, ever since he had watched a male wasp take up the same position on a female wasp during sexual intercourse. Here was proof.

Pouyanne was the first to find the answer to the question which had puzzled many observers before: why did the orchid flowers look like wasps? The resemblance was intended to deceive male wasps into believing that the flowers were female wasps. What the judge was witnessing was an attempt by the wasp to mate with the flower; the whole thing was an elaborate fraud designed to use the wasp's powerful sex drive in order to cross-pollinate the flowers and so produce a new generation of orchids.

In later chapters we will study more closely the various forms of sexual deceit practiced by plants, and some of the methods by which they use not only wasps and bees but birds, butterflies, moths, flies and even slugs to fulfill their sexual needs.

Judge Pouyanne's discovery was perhaps the last great breakthrough in our knowledge of plant sex, and like all the previous discoveries in the field it was greeted with a mixture of prudery, disbelief and apathy. True, in 1916, when he published his findings, the French were engaged in a bitter and bloody war; the serious business of the day was violence, not sex. But in addition to that Pouyanne was an amateur, and botanists – like members of any other profession – do not like an outsider to point out things that they ought to have noticed themselves. So for several years they ignored the judge's discovery; after all, had not the great Charles Darwin himself, after having his attention drawn by a Mr. Price to "attacks" by bees on similar insect-like orchids, stated in writing that he could not possibly conjecture what it all meant?

Others dismissed the judge's discovery as the fantasy of a dirty old man. Some people, they said, would see sex in anything; it was all part of this modern obsession with sexual matters, and the judge's suggestive remarks should never have been printed in a respectable journal devoted to the pure and wholesome subject of horticulture.

So Pouyanne's discovery was shrugged off and ignored by

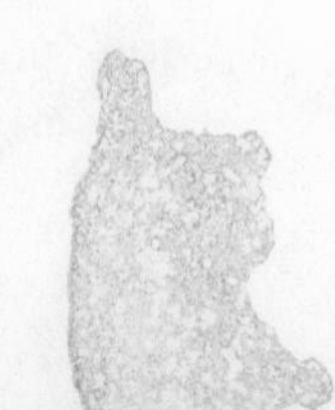

most professional botanists for many years. Indeed, in 1926, just ten years after the publication of Pouyanne's findings, the Viennese naturalist Raoul Heinrich Francé was still managing to get hold of the wrong end of the stick when he quoted a botanical expert as saying "No wonder that insects neglect the *Ophrys* flowers! The design of the lip really helps to scare away unbidden guests. The flower wants no visitors, therefore it mimics the form of a visiting insect." Which only shows that if you shut your eyes and your mind to the sexual facts you finish up with the precise opposite of the truth: what was intended by the flower to say "Come and make love to me" is misunderstood as saying "Keep off."

One or two keen observers such as Col. M. J. Godfery were open-minded enough to look into Pouyanne's findings and express their belief both in the correctness of his observations and in the conclusions he drew from them. But it was not till nearly half a century later that he was proved right in every detail, through the painstaking work of the Swedish biologist Dr. Bertil Kullenberg, who studied the whole thing from the scientific point of view – botanical, zoological and chemical. Kullenberg published his findings in a classic book, *Studies in Ophrys Pollination*, which has been described as reading like a novel, combining as it does just the right mixture of mystery, suspense, detection and sex.

Most discoveries in the field of plant sex (and for that matter animal sex, particularly that of the human animal) have been greeted in much the same way, with the extreme hostility of both the academic and the ecclesiastical establishments – the former because such discoveries disturbed the existing body of learning, and the latter because in every culture the subject of sex has tended to be frowned upon by the religious authorities, recruited as they are largely from those who disapprove of the flesh and find the facts of life distasteful.

Even today, when a woman loses her virginity she is said to have been "deflowered," the suggestion being that she no longer has the sexless quality of a flower. That attitude goes right back to the legends of the ancient world, in which plants seem to have been thought of not only as sexless but as the final refuge from sex, with all its passions and problems. When the

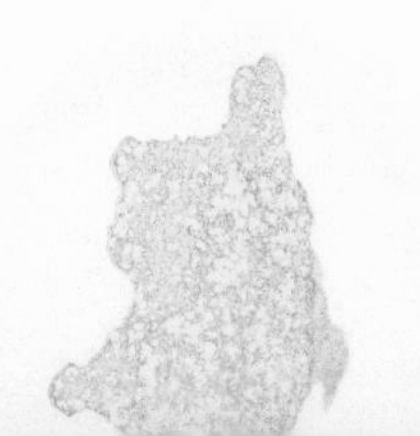

poet Ovid published his sex manual *The Art of Love* (described in the Encyclopaedia Britannica as "perhaps the most immoral work ever written by a man of genius"), he shocked the establishment of his day, headed by the Emperor Augustus, who later had him banished during a Clean-Up-Rome campaign. Ovid tried to change his image from one of lewdness to one of respectability by writing the *Metamorphoses.* To show that he was really a decent god-fearing – or rather gods-fearing – man, he collected together the traditional Greek and Roman legends concerning miraculous transformations of gods and goddesses, mortal men and women, animals and plants. In every story in the poem in which plants appeared they represented "purity," which to traditionalists then – as always – meant sexlessness.

We read of Narcissus, who because of physical or psychological impotence refused the advances of the nymph Echo, fell in love instead with his own reflection, and turned into the flower that bears his name. From that legend psychologists coined the word narcissism, defined in psychoanalytical dictionaries as "*A sexual perversion in which the sufferer's preferred object is his own body; a defense mechanism resulting from the introjection of a denied love-object.*" Ordinary non-psychologists call it infatuation with oneself. To the sufferer, however, it is not mere infatuation, but the real thing.

Then there was Daphne, to whom Apollo, the sun-god, took a fancy. According to Ovid, Apollo was set on fire by her beauty, so that "his whole heart was aflame." Frightened by his lecherous looks and his improper suggestions, Daphne ran away "as the lamb flees the wolf, or the deer the lion, or as doves fly from an eagle." Apollo ran after her. She "felt his hot breath touch the hairs on the back of her neck." Terrified, the maiden cried out to the river-god Peneus, her father: "Destroy my beauty that makes me so attractive." So Peneus turned her into a laurel tree and saved her virginity for ever.

The god Pan, notorious for his never-satisfied lust, fell for the water nymph Syrinx and, being a coarse character who knew nothing of the arts of foreplay and seduction, tried to rape her. She prayed to the heavens to be saved and just in the nick of time was turned into a clump of reeds. Cheated of straightforward sex, Pan made a pipe out of the reeds and went around playing tunes on the remains of the nymph he desired: perhaps this was the first recorded instance of a kind of necrophily.

Hyacinthus was a very good-looking boy whose beauty aroused the desires of Apollo (who seems to have been both promiscuous and bisexual in his tastes). Zephyrus, the god of the west wind, had also taken a fancy to Hyacinthus, and determined that, if he could not have the boy, Apollo should not have him either. So when Apollo threw a discus into the air during games, Zephyrus blew it so that it struck the boy on the head and killed him. Apollo was so upset at what he thought to be the results of his own clumsiness that he changed Hyacinthus' spilt blood into the flower which has borne his name ever since. So Hyacinthus retained his unsullied purity in the form of a plant; he may have been killed in the process, but at least he had not suffered what moralists in later times were to call "the fate worse than death."

Nowadays we might assume that Ovid had not intended his collection of tales in the *Metamorphoses* to be taken literally but to be read as a fanciful expression of poetic feelings about the world of nature. There is no doubt, however, that he presented the poem as a factual record, because he ended by describing the metamorphosis of Julius Caesar into a star and promising immortality to his successor Augustus. Sad to tell, Augustus was not impressed by such an obvious attempt to gain his favor, and Ovid died in exile.

For the whole of the three hundred years before Ovid wrote the *Metamorphoses*, the study of plants had been entirely dominated by the work of the Lesbian scholar (that is, from the isle of Lesbos) Theophrastus, who became perhaps the most important teacher in Greece after succeeding Aristotle at the Lyceum, and who presided over the leading Peripatetic school for thirty-five years. The two gigantic works on the vegetable kingdom by Theophrastus, *On the History of Plants*, in ten books, and *On the Causes of Plants*, in eight books, were without question the most influential writings on botany of their time, and accepted as the unchallengeable authority on the subject for nearly two thousand years, from the third century B.C. to the sixteenth century A.D. Theophrastus prudently avoided calling the old myths into question. After all, it was only a few years before his birth that the great philosopher Socrates had been sentenced to death for, among other things, failing to show proper respect towards the gods and for having new ideas about religion. The real reasons why he was convicted may have been more complex and political, but the traditionalists had won a victory, and anyone unwise enough at that time to question the literal truth of the old legends might well have found himself accused of blasphemy.

It is not surprising, therefore, that nowhere in any of the volumes of his botanical works does Theophrastus seriously challenge the accepted belief in the sexless nature of plants. There are a few sentences in his observations about palm trees which suggest to some scholars that he may have had an inkling of plant sex, but they are put in such delicate and vague language that they did nothing to generate either excitement or criticism.

In any case, when Theophrastus was making his observations about plants there was no possibility of getting at the true facts of life in a sexual sense. The central principle that the female plays at least as important a part as the male was quite beyond the grasp of the human mind and would remain so for the next two thousand years, till the invention of the microscope and the discovery of cells. Even in the animal kingdom, where sex was too obvious a fact to be ignored, the accepted doctrine was that the female contributed nothing at all except accommodation; she was a mere receptacle, in which the male deposited his "seed," containing a fully formed embryo which needed nothing from the mother but warmth, shelter and nourishment to enable it to grow.

The lesser importance of the female was an article of faith in the Peripatetic school, having been stated very clearly by Aristotle himself. In doing so he was repeating the accepted dogma expressed with the authority of a god on the Athenian stage little more than a century before Theophrastus wrote his botanical works. Aeschylus, the father of Greek drama, in his play *The Eumenides* has the god Apollo defend Orestes against a charge of murder for killing his mother. In a key speech Apollo states as unquestionable fact "The mother is not parent of her so-called child but only nurse of the new-sown seed. The man who puts it there is parent; she merely cultivates the shoot, host for a guest . . ."

Just over three hundred years after the death of Theophrastus the Roman naturalist Pliny the Elder showed some glimmerings of a knowledge of plant sex in his encyclopedic work *Natural History*, published in 79 A.D. Dealing with palm trees, Pliny plagiarized, and elaborated in a somewhat coy manner, the suggestion made by Theophrastus and already mentioned. In an English translation published by Robert John Thornton, M.D., in his work on plant sex *The Temple of Flora*, published in 1807, the passage from Pliny read as follows: "it is confirmed that the wild *female palms* do not produce fruit without the assistance of the *male*, and for this purpose the *females* bend their boughs to him for mutual embrace. He also marries with the other *female palms* by his gentle sighings, tender looks, and the dispersal of a *powder*. This *male tree* being cut down, the widowed females afterwards become sterile. This *love* in plants has been observed by men, who imitate it by the scattering of flowers and down of the male, or even by the dispersion of the *powder*, upon the *females*." Not for the first time, or probably the last, it seems that practical gardeners, in touch with nature, used the facts of life as they found them to increase fruitfulness without bothering their heads with the theories of botanists, philosophers or other intellectuals.

In spite of that small chink in the accepted doctrine of the sexlessness of plants, those words by Pliny seem to have been the last on the subject till modern times. Too many traditions would have been undermined if such new thinking was allowed to go unchecked. Not only philosophies but whole religions had been founded on the denial of the facts of sex, particularly of the main one that the female played an equal role with the male. Sacred traditions showed that on more than one occasion gods had been able to beget offspring in their own likeness, untainted with human imperfections, by impregnating a female mortal chosen by them as a suitable vessel. How could the offspring remain uncontaminated by human faults if the female, instead of being simply a receptacle, had played as important a role as her supernatural seducer and had contributed some of her own human genes to the infant?

Perhaps the male-dominated establishment realized, consciously or unconsciously, that once it was admitted that the mother shared in the creation of the offspring it would follow inescapably that she was *more* important than the father, because she contributed not only board and lodging but half the infant as well. (In fact, the female – whether animal or plant – contributes a good deal more than half the offspring, as we shall see later.)

For the next fifteen hundred years after Pliny's work, all through the decline of belief in the old gods, the fall of Roman civilization and the rise of Christianity, nothing more appears to have been observed, or at any rate written, on the subject of plant sex. Many volumes, full of descriptions of plants and their uses, were produced by the monks who kept some kind of learning alive during the Dark Ages, and who studied the vegetable kingdom with some thoroughness in their search for herbal remedies. But people who took up the monastic life were less likely than most to be receptive to the notion of sexual activities being carried on incessantly, all around them, by every form of life; it was to get away from that sort of thing that many of them had retreated from the world.

So the idea that plants, those pure and innocent things which God had created for man's use and contemplation in the Garden of Eden before the Fall (itself all bound up with sex), should open their flowers and show them off not just to delight the eye but as a brazen way of exposing themselves and inviting sexual advances would have horrified the faithful and caused them to do penance for entertaining evil thoughts.

When at last the discovery of plant sex was made, just three centuries ago, it was not made by the monks and scholastics, whose existence was one of pious contemplation, who relied for authority not on observation of the real world but on tradition, and many of whom were uneasy about the facts of life. The breakthrough came from medical men, whose interest in plants was the very practical one of using them to treat disease, and whose anatomical studies had given them exactly the right training and outlook.

The sixteenth century saw a tremendous upsurge of interest

in the structure and functioning of the human body. Stimulated by the anatomical drawings of the new wave of renaissance artists such as Leonardo da Vinci and by anatomists such as Andreas Vesalius, physicians began to examine real human beings instead of accepting the authority of ancient philosophers. Often at considerable personal risk from superstitious bigots, they cut up corpses to find out what had made them work. The causes of this sudden upsurge of enquiry and activity were many and varied, including the exploration of new lands and the discovery of new peoples, new plants and new drugs.

But perhaps the biggest single cause was the introduction of syphilis by the returning explorers and their crews. By the end of the fifteenth century the disease had become epidemic all over Europe, causing terrible misery and loss of life. It has been said that no royal or noble family – that is, no family that could afford the services of a physician – was free of the "pox," as it was soon called. Something had to be done, and done quickly. As a result, knowledge of infectious disease increased rapidly as rich men paid large sums of money so that methods of treatment could be found. The manner of transmission through sexual intercourse was established, and this in itself gave great impetus to the study of sex and the sexual organs of men and women. To tackle the seriousness of the epidemic, false modesty had to be abandoned, and physicians began to look at the private parts of even their female patients. Soon male doctors were actually attending at childbirth, much to the outrage of the midwives, whose ignorance, superstition and just plain dirt had kept up the high rate of mortality in medieval childbirth.

It is perhaps fitting that the first epidemic of sexually transmitted disease was to lead so rapidly to a vast increase in the knowledge of the human body and the discovery of the facts of sex among humans, animals and, shortly afterwards, plants.

By the seventeenth century the medical world was in a ferment of new ideas and discoveries. In 1628 the English physician William Harvey had published his treatise on the circulation of the blood, which in spite of resistance from the establishment quickly overthrew the ancient beliefs in "spirits" asserted by Aristotle and the Greek physician Galen on which medieval medicine had been based. The authority of old writings was being challenged by observations of the real world.

Twenty-three years later, in 1651, Harvey published his *Treatise on Generation*, which startled the world by setting out the true facts of sex. *Omne vivum ex ovo* was the principle he stated: in other words, to quote from his work, "All living things, even those animals which bring forth their young alive, and man himself, are produced from eggs."

Those were the words which ushered in the modern revolution in our study of sex, and finally swept away the belief, unquestioned up to then, that any form of living creature could arise by a process of spontaneous generation. Life was not suddenly created; it was made, from the fertilization of an egg.

Once the real nature of the sexual process among animals had been established, it was not long before the same discovery was made in the world of plants. Marcello Malpighi, an Italian doctor and lecturer in medicine at Bologna, was one of the first to use the microscope to study animal structures, and soon he became curious about the structure of plants as well. In 1672 he published his *Anatomia Plantarum*, which recorded his detailed and painstaking observations on plant anatomy and physiology. In the same book he included his *Observationes de ovo incubato*, which gave an accurate account, with excellent illustrations, of the development of the chick in the egg. Yet he did not make the final mental connection that the same sexual process that made the chick went on in plants too.

Four years later the idea of plant sex was finally suggested by an English physician, Sir Thomas Millington, F.R.S., who was later to become President of the Royal College of Physicians. Since the Royal Society, of which he was an eminent member, was in correspondence with Malpighi, it is possible that Millington had seen a copy of *Anatomia Plantarum* and put two and two together to connect plants and sex. He did not, however, make his idea public but simply mentioned it to a fellow physician and botanist Nehemiah Grew, who was quick to see its possibilities and caused quite a stir when he included the suggestion of plant sex in a lecture, *The Anatomy of Flowers,* given by him to the Royal Society in 1676. The lecture created so much interest that Grew included it in his *Anatomy of Plants*, published in 1682. He gave full credit to Millington, who he said "suggested to me that the attire (that is, the stamens) doth serve as the male for the generation of the seed."

In 1686 the great John Ray, who was born the son of a village blacksmith and became known as the father of English natural history, published his *Historia Plantarum* in which he gave a favorable view of Millington's idea as expressed by Grew. Ray added cautiously "This opinion of Grew, however, of the use of the pollen wants more decided proofs; we can only admit the doctrine as extremely probable."

Eight years later the proofs were provided by a German named Rudolf Jakob Camerarius who, besides being Professor of Medicine at Tübingen, became Director of the Tübingen Botanical Gardens. His *Epistola de Sexu Plantarum*, dated 25th August 1694 and addressed to his friend Valentini, Professor of Medicine at Giessen, is the first published evidence on the subject, and is generally accepted as establishing plant sexuality for the first time. By careful and painstaking experiments, such as removing the stamens from male flowers of the castor-oil plant and showing that the female flowers then failed to set seed, and also by showing that no seed was developed after the removal of the female "tassels" of maize plants, Camerarius placed the facts that plants have sex beyond doubt. On the question of bisexual flowers, he quoted the discovery by Swammerdam that snails were hermaphrodites, and made the point that while bisexuality was evidently uncommon in the animal kingdom it may well be the usual thing in the plant world. He did, however, rather too easily assume that hermaphrodite flowers pollinated themselves. The idea that the male and female parts of the same flower could become sexually active at different times, so that the flower in effect changed sex, and therefore could only fertilize, and be fertilized by, other flowers, does not seem to have occurred to him.

Though the publication by Camerarius of his conclusions in 1694 represents perhaps the most important date in the history of human acceptance of plant sex, it did not revolutionize scientific thought overnight. Many people refused to accept his findings. Many others remained ignorant of them; dissemination of the results of scientific discovery was not as complete or as quick then as it is nowadays. Even the eminent French botanist Tournefort, in 1700, six years after the publication by Camerarius, still stated his belief that the purpose of stamens was to get rid of surplus sap by excreting it in the form of pollen. It has always been the mark of the sexually unenlightened to get the excretory functions confused with the sexual ones; that is perhaps the chief reason why they so often call sex filthy.

A few botanists, excited by the findings of Camerarius, were stimulated to try some experiments themselves. The English "natural philosopher" Richard Bradley, who not only became Professor of Botany at Cambridge but was a well known popular writer, published in 1717 a work entitled *New Improvements of Planting and Gardening.* In it he described how he carried out some sexual investigations on tulips in his garden by cutting off the male organs of twelve of them and leaving the rest – about four hundred plants some distance away – with their full male equipment. The result was that, in his words, "the twelve being thus *castrated*, bore no seed that summer, while on the other hand, every one of the four hundred plants which I had let alone produced seed."

That seems to have been the first sexual experiment performed on hermaphrodite flowers. Bradley also discusses the removal of male catkins from a nut tree which is then barren unless the female flowers are impregnated with pollen from catkins taken from another tree, "gather'd fresh every Morning for three or four Days successively and dusted lightly over the female parts." It appears that Bradley was discovering not only birth control but artificial insemination; he includes an account of the creation of a new man-made cross between a carnation and a sweet william by Thomas Fairchild, and looks forward to many more artificial hybrids in the future, in order to breed better plants.

In 1721, four years after Bradley's publication, Philip Miller repeated the experiment by castrating twelve more tulips, but this time he noticed something quite different. All the castrated flowers developed seed. The reason was given by him in a letter to his friend Patrick Blair, a surgeon, who published the explanation in the *Philosophical Transactions of the Royal Society*, of which he was a Fellow, under the title *Observations upon the generation of plants.* About two days after the flowers had been castrated, Miller noticed "Bees working on Tulips in a bed where he did not take out the *Stamina*, and when they came out they were loaded with Dust on their Bodies and Legs; he saw them fly into the Tulips where he had taken out the *Stamina*, and when they came out he went and found they had left behind them sufficient to impregnate these Flowers, for they bore good ripe seed – which persuades him that the *Farina* may be carried from place to place by insects." (Farina, or in full *Farina fecundens* – which means fertilizing powder – was the name then usually given by scholars to what we now call pollen.)

What Philip Miller had discovered was the part played by bees in the pollination of plants. It was now at least established that hermaphrodite flowers were not necessarily self-fertilized. Soon it was being argued that flowers never fertilized themselves, and this view was held for many years, apparently not merely on scientific grounds but on moral ones, as a condemnation of what was seen as the sin of incest.

A few years after Miller's observation that insects could cross-pollinate flowers, news came from America that wind could do so too. James Logan, who was Chief Justice and President of the Council of Pennsylvania, described in a letter dated 20th November 1735 to his friend Peter Collinson, a Fellow of the Royal Society, some experiments he had made on plants of "Mayze, or *Indian Corn*" in his garden in Philadelphia. From some plants he had cut off the male flowers at the top before they produced any pollen; from others he had removed the whole female tassels, and from others still he had removed a quarter, half or three quarters of the tassel. With yet other plants he had enclosed the female parts, just before the tassel (or silk, as he called it) appeared, in muslin bag: "the Fuzziest or most Nappy I could find, to prevent the passage of the *Farina*, but that would obstruct neither the Sun, Air or Rain. I fastened it also very loosely, so as not to give the least check to vegetation."

The cobs in the muslin bags produced no seed; nor did those which had had the whole tassel removed. The ones that had had part of the tassel removed produced seed only from those grains whose silk had been left intact. The plants which had had their male flowers removed produced no seed at all, with one exception. A large cob on one of the castrated plants produced about twenty seeds out of a possible five hundred or so. The reason, James Logan suggested, was that the cob in question had had its female tassel exposed in the direction of some uncastrated plants a short distance away, from which the wind must have carried a certain amount of pollen.

In the three centuries or so that have passed since Camerarius first proved beyond doubt the existence of plant sex, a vast amount of information on the subject has been collected by means of countless experiments and observations. The process still continues. But the basic facts – the existence of the male and female genital organs, their structure and function, their methods of operation and the ways in which they manage to achieve pregnancy, with or without the help of outside agents such as insects and wind – were all established within the first thirty or forty years after the pioneering work of Camerarius, using much the same experimental techniques as the ones he had started.

Professors K. Faegri and L. van der Pijl, in their important book on the subject *The Principles of Pollination Ecology*, suggest the reason for this rapid series of discoveries in the following words: "Perhaps it is more than a coincidence that the sexually repressed Middle Ages could not grasp an idea which presented no difficulty to the more licentious eighteenth century."

Later we will take a look at some of the most interesting and significant discoveries of more recent times. But before that it is necessary to consider the man who did more than anyone else, before or since, to put the whole idea of plant sex on the map.

The great Swedish botanist, Carl von Linné, better known as Linnaeus, was born in 1707, just thirteen years after the breakthrough by Camerarius. The son of a pastor, young Carl was expected to study theology, the most fashionable and highly esteemed subject of the day. However, his unsatisfactory performance at the Gymnasium of Vaxio, where he was sent as a pupil, caused the authorities to decide that he was "not sufficiently skillful" for theology, so the study of medicine was chosen instead. His mother, who was only eighteen when he was born and whose relationship with him seems always to have been strained, made no attempt to hide her disappointment and disgust. Medicine was not highly thought of as a profession in rural Smaland, being considered work for a technician rather than an intellectual.

In view of his later achievements, it seems hardly likely that Linnaeus could have lacked the mental ability to study theology. Indeed, his epoch-making work in botanical classification shows all the marks of a disciplined theological mind; some present-day botanists would say that it was too disciplined and too theological.

The truth seems likely to be that young Carl let his attention wander from the study of theology to the subject of sex.

To most young men that would have meant the awakening of interest in girls, but Carl's youthful interest in sex seems to have been directed towards plants. He had formed a warm attachment at the gymnasium to his teacher Rothman, who was absorbed by the new discoveries concerning sex among the flowers and aroused his young student's interest in the subject by showing him the available reports, including a paper which had greatly stimulated him: Vaillant's *Sermo de Structura Florum*. This, published in 1718 when Carl was entering his teens, gave most of the then established facts in a readily digestible form and must have helped to kindle that interest which was to grow into an obsession during Carl's adolescence and early manhood.

In view of the hostility which the dicoverers of plant sex met with during their lifetime, it is ironic that nowadays nations should compete with each other for the honor of being first in the field. The French have put up a notice beside a pistachio tree in the Jardin des Plantes in Paris claiming that this was the very tree that enabled Vaillant "to discover sexuality in the vegetable kingdom in 1716"; in fact that was twenty-two years after the publication by the German Camerarius and forty years after the disclosure by the Englishmen Millington and Grew.

At the age of twenty-two, after entering Uppsala university as a medical student, Linnaeus put his ideas on the subject down on paper in an essay entitled *Praeludia sponsaliorum plantarum.* It was the custom of the time for all learned papers to be written in Latin, and Carl was particularly fluent in that language, having been taught by his father not only to write but even to converse in it from a very early age. The title of the essay is usually translated as "Preview of the betrothal of plants," but the literal meaning of *Praeludia* is the sexually more appropriate word "Foreplay." The text, embellished with line drawings by the author, is rather fanciful and somewhat juvenile, but it already contains many of the ideas which were to make Linnaeus famous and on which he was to found the system of plant naming and classification that botanists use to this day.

The essay so caught the fancy of Olof Rudbeck, professor of botany at the University, that he appointed Carl, young as he was, to demonstrate the subject to his fellow medical students, using the plants in the botanic garden as illustrations. These demonstrations were probably the first example of sex education to make use of the bees and the flowers to explain the facts of life.

The subject of sex among the plants increasingly fascinated Linnaeus, and his medical studies were interrupted by journeys to Lapland and other places to examine the flora. In 1734 a teacher named Browallius advised him to do two things: marry a rich girl and finish his studies abroad. Linnaeus quickly found a suitable girl, a doctor's daughter aged eighteen named Sara Lisa Moraea, and asked for her hand in marriage. Her father consented on two conditions: that there should be no wedding for three years and that meanwhile Carl must get his medical degree. In 1735 Linnaeus went to Holland, where he very soon gained his degree at the University of Harderwijk. He had taken with him from Sweden several of his manuscripts elaborating the ideas suggested in his "Foreplay," and setting out a new method of classifying plants according to their sexual characteristics. The men of science to whom he showed his work at once realized that it was of major importance and with the help of rich patrons arranged for its publication.

Within the three years after his arrival in Holland Linnaeus had no fewer than fourteen printed works to his credit. When he left in 1738 he was internationally famous in scientific circles as the founder of modern botanical classification. His publication of earliest date, December 1735, is *Systema naturae*. In it Linnaeus set out the principles of his sexual system of classification, a system which he followed in all his later works. The key to the vegetable kingdom given in the *Systema* is intended to enable any group of plants to be quickly identified according to the number and arrangement of the sexual parts of the flower. The book was immensely successful; by 1759 it had gone through ten editions. It took the fancy not only of the scientist but the more general reader, perhaps because some of the language Linnaeus used seemed to attribute to plants the same sort of sexual attitudes and behavior as those shown by human beings.

He describes one group of plants in terms that, translated from the Latin, mean "openly celebrating marriage in a way that is obvious to all." Another group is described as having marriage "secretly arranged." The first section is divided into those with flowers in which "husbands and wives enjoy one and the same marriage bed" and those with flowers in which "husbands and females enjoy separate beds." Note the distinction Linnaeus draws between "wives" (in his original Latin test *Uxores*) in the first case and females *(Feminae)* in the second; there is a strong suggestion that the latter are "kept women" in separate establishments.

Linnaeus had brought great honor and distinction to Holland by his publications. He was begged to stay, but he refused. As tidy-minded in his life as in his work, he saw no reason why success should change the routine he had arranged before leaving home. He had agreed to three years, the three years were up, and Sara Lisa was waiting for him back in Sweden. In May 1738 he left Holland, in September he set up in practice as a doctor in Stockholm and the following year he was married. With his mania for classifying everything he even classified his bride by sexual category. "My monandrian lily," he called her, the choice of flower signifying virginity and the word monandrian meaning "having only one man." In more practical terms he classified her by wealth; in a letter to a friend he wrote "I have just married the woman I have wanted to marry for years and who is, between ourselves, quite rich."

The contribution by Linnaeus to our methods of naming and classifying plants was far-reaching. When he started his work, the whole subject was in complete chaos, plants being given names sometimes forty or fifty words long, so that it was impossible to list them, to identify them or to arrange them according to their relationships with each other. Several botanists had attempted to classify plants in quite arbitrary ways, by such things as whether they were trees or herbs, grew on dry land or in wet places, or were poisonous or edible.

Sex was the thing that enabled Linnaeus for the first time to devise what he called a "natural system." In his long work *Philosophia botanica* he even indulged his taste for classification by classifying other botanists. Some were orthodox; they were the correct ones who arranged plants according to the "fructification," which meant those parts – flower, fruit, seed – designed to beget a new generation: in other words the genital organs. Others were heterodox; they were the misguided ones who arranged plants by any other criterion: the alphabetics (by alphabetical order), the rhizotomi (by root structure), the phyllophiles (by leaf-shape), the physiognomists (by external appearance) and so on. Even among the ranks of the orthodox all the other botanists were wrong: fructists (classifying by fruit), corollists (by petals) and calicists (by sepals). Those in the final category, the sexualists, were the only really correct ones. They classified entirely on the basis of the sexual organs. (*"Ut ego,"* wrote Linnaeus, meaning "like me.")

In fact the Linnaean system has had to be very much modified over the years. In spite of his claim that he was the first "sexualist" among plant classifiers, Linnaeus was only partly so. In particular he overstressed the importance of the male organs; he based his arrangement of flowering plants first and foremost on the number of stamens present, their relative lengths, and whether they are separate ("husbands not related to each other") or joined together ("husbands tied in bonds of brotherhood"). True, Linnaeus does bring the female organs into some of his classifications, but in a strictly subordinate role. Nowadays he would certainly be called a male chauvinist.

It is interesting to note that during the growth of the feminist movement, more than a century and a half after Linnaeus published his *Systema naturae*, the opposite point of view was expressed. In 1906, there appeared a small book on the elements of botany entitled *The Study of Plant Life*. It had considerable success as a text book in schools, and nowhere in its pages does the word "sex" appear. However, in the chapter headed "Flowers" we find these words: *"Now we have come to the heart of the flower, and find there the most important thing in it . . .* a tiny green structure very like a pea pod with a little sticky knob at the tip." That female chauvinist assertion that the most important parts of the flower were the female organs was written by a woman of 26, the youngest Doctor of Science in Britain and lecturer in palaeobotany at the University of Manchester. Her name was Marie Stopes and she was later to become famous – or notorious, according to one's point of view – as the author of the bestselling sex manual *Married Love*, pioneer of birth control and fighter for the rights of women.

In this brief history of the recognition of plant sexuality it is necessary to dwell at some length on the work of Linnaeus, not only because he was more influential than anyone else in getting the idea of classification on sexual grounds accepted but because in addition to his scientific publications he wrote on the subject for a wider audience of non-specialists. His *Prize Dissertation on the Sexes of Plants*, published in 1759, reminded the world that he could claim no personal credit for discovering the sexual facts, which had first been glimpsed eighty-three years before. "The English report," he wrote, "that their *Millington* was the *first true discoverer of this doctrine* . . . They contend that about the year 1676 he saw the whole mystery; and, in truth, not long after, Grew and Ray, both Englishmen, explained the matter further."

It was, however, in vain that Linnaeus pointed out that he had discovered nothing, but merely used other people's discoveries to construct better methods for the recording, arranging and naming of plants. Because he had caught the world's attention as the man who proposed setting up a new system based on sex, he was made the target of all the abuse that leading figures of the establishment always let fly whenever the word "sex" is used.

The Church could be relied upon to make a fool of itself over the issue, and so it did. Some clerics called for the works of Linnaeus to be banned. An English clergyman, the Reverend

Samuel Goodenough, later to be appointed Bishop of Carlisle, was so shocked and horrified by the very idea of plants' genital organs that he wrote in a letter "nothing could equal the gross prurience of Linnaeus' mind."

Goethe, greatest of German poets, expressed the view that the introduction of Linnaeus' sexual system was likely to cause a great deal of embarrassment in the teaching of botany to chaste young people. He had a considerable knowledge of the natural sciences and wrote an important botanical work which attempted to explain the changes undergone by plants during their life cycle. Perhaps it was his reservations about the Linnaean system rather than genuine feelings of embarrassment that made him say what he did. After all, from the records we have of his own series of love affairs he seems not to have suffered excessively from sexual inhibitions himself.

Even some leading members of the scientific and literary establishments protested noisily at Linnaeus' "indecency." In France his work was ridiculed by Buffon; in Germany it was attacked by Heister (who seems to have been actuated largely by envy); in Switzerland it was rejected by Haller; in Italy it met with the heavy-handed opposition of Pontedera; in England it was greeted with sarcastic spite by Alston; and in Russia it was violently denounced by the botanist Sigesbeck, on the grounds of its disgusting obscenity. (So viciously and irrationally did Sigesbeck continue to attack the sexual system after it had been generally accepted that he was finally dismissed from his post as professor of botany at the Imperial Academy at Petersburg. He asked Linnaeus to help him get a job as a head gardener. Linnaeus, never one to triumph over a defeated enemy, named a plant *Sigesbeckia*, in honor of the fallen professor.)

Although our whole method of naming plants on the binomial principle dates from Linnaeus' *Species plantarum*, published in 1753, many present day botanists now believe that his method of plant classification actually delayed botanical progress. Because of the narrowly religious background of his childhood he had a fundamentalist belief in the dogma (somewhat old-fashioned even at the time) that in the beginning God had created a fixed number of species and genera, and that this number had remained the same ever since. Added to this, he believed that, since scripture tells us man was made in the image of God, any classification should start with the highest organism (the "ideal") and work its way downwards to lower forms of life. Since few would argue that lower forms had come from higher ones, this method of arrangement effectively blocked the notion of biological time; together with the dogma of a fixed number of species this must have considerably delayed the possibility of the idea of evolution.

In addition, his training in scholastic traditions and in formal logic on Thomistic-Aristotelian lines caused Linnaeus to think in abstract terms, so that all too often what he called his "natural system" became highly artificial, with plants being forced to fit into categories instead of categories being developed to fit the plants.

Even the meticulous tidy-mindedness which had enabled Linnaeus to devise his classifications, perhaps originally an unconscious method of trying to gain favor with his unloving and disapproving mother, was to prove his undoing in the end. He became increasingly obsessive during the last fifteen years of his life and died in 1778 at the age of 70 in a state of dementia, repeatedly expressing terror at the thought of *Nemesis divina*, the "divine retaliation," which was in store for him. The source of his guilt feelings can only be guessed at now; perhaps it had something to do with the accusations of obscenity that had been hurled at him for introducing the subject of sex into systematic botany.

In 1750 Arthur Dobbs made some careful observations on the bees around his home in Northern Ireland. His findings were published in the *Philosophical Transactions of the Royal Society*. They confirmed the observations by Philip Miller that flowers may be pollinated by insects. In addition they confirmed a comment made in ancient times by Aristotle in his *History of Animals* that a bee when foraging keeps to one sort of flower "and never meddles with another kind until it has returned to the hive." Dobbs, after establishing the "faithfulness" of a bee to its chosen flower, adds "Now if the Facts are so, and my Observations true, I think that Providence has appointed the Bee to be very instrumental in promoting the Increase of Vegetables."

The first botanist to perform large-scale experiments in hybridization was Joseph Gottlieb Koelreuter, who during the six years from 1761 to 1766 published descriptions in great detail of both the male and the female parts of flowers, showed different types of cross-pollination, and explained what nectar was and what it was for, to reward the visiting insects for their labor in the cause of sexual reproduction.

The next important observations were made by Christian Konrad Sprengel in his great work *The Secrets of Nature Revealed in the Structure and Fertilization of Flowers.* He dealt in great detail with some 500 species of plant and described with painstaking accuracy the adaptations of their sexual parts. It was Sprengel who pointed out the very widespread occurrence of a "sex change" in flowers from male to female, in which the anthers shed their pollen before the stigma is sexually ready, and so avoid self-fertilization. He was also the first to discover the opposite condition, in which the female part of the flower reaches sexual puberty before the male.

Sprengel was a first-class observer and described in great detail the amazing variety of ways in which flowers adapt themselves to pollination by all sorts of insects: flying, crawling and hopping. He also examined wind-pollinated flowers, pointed out that they had to produce pollen on a prodigal scale and showed why the male anthers exposed themselves as fully as they did and why the female stigmas were so large and often feathery, so as to trap as much pollen as possible before the wind blew it away. From his observations he stated that "nature seems unwilling that any flower should be fertilized by its own pollen."

Shortly afterwards Thomas Knight, president for many years of the Horticultural Society of London, made some experiments on cross-pollinating peas and came to the conclusion that cross-fertilization was of great benefit; among the offspring from his artificial cross-mating were "a numerous variety of new kinds produced, many of which were, in size and in every other respect, much superior to the original kind and grew with excessive luxuriance."

The next vital contribution to the subject of the fertilization of flowers came from Charles Darwin. Though Darwin is chiefly remembered in the public mind for a statement he never made, that men were descended from apes, a large proportion of the work on which he based his epoch-making theory of evolution was carried out among plants, not animals. Perhaps he had been drawn to the subject by the works of his illustrious grandfather Dr. Erasmus Darwin, who had been greatly excited as a young man by the publications of Linnaeus and his followers on the sexual system and had written a somewhat romantic and mildly salacious poem on the subject entitled *The Loves of the Plants*, some lines from which appear at the beginning of this chapter. Erasmus Darwin published many other works, including *Phytologia, or the Philosophy of Agriculture and Gardening*, which caused a stir in 1799 because of its claim that plants can not only feel sensations but exercise free will; so Charles came of a family that had already had its share of public controversy.

The howls of rage that greeted the publication of Charles Darwin's theory of evolution in 1859, exactly a century to the very year after the appearance of Linnaeus' *Prize Dissertation on the Sexes of Plants*, made any attacks on Linnaeus seem tame and restrained by contrast. Not only was sex once again rearing its head, but Darwin had committed the unforgivable sin of suggesting – against the assertions not only of the Bible but of Linnaeus himself – that there was not a fixed and unchangeable number of species created by God in the beginning but that new forms were constantly evolving.

Darwin's work was thunderously denounced from pulpits throughout the world as blasphemy. A succulent new topic had appeared with which to harangue the faithful and castigate the wicked, with that combination of moral indignation and total ignorance of which powerful sermons are made. The echoes have still not quite died down; a few clergymen – from laziness, lack of subject matter or even in extreme cases religious conviction – sometimes to this day dust off the old anti-evolution stuff and give it an airing in front of their congregations. But they are a dying breed.

Why was Darwin's work subjected to such extravagant abuse? It could hardly have been the title of his book, which went out of its way to be serious and sober to the point of ponderousness, *On the Origin of Species by Means of Natural Selection, or the Pre-*

servation of Favoured Races in the Struggle for Life: hardly a snappy or controversial looking title even by the standards of 1859, when scientists were expected to show a certain gravity of style in their offerings.

Why the storm then? Partly it must once again have been caused by distrust of any attempt to shine new light into dark corners. To some extent this is simple job protection, because those whose living depends on dogma feel threatened when that dogma is undermined (and people who have served a dogma for a long time probably cannot be retrained for more useful work). Then there is a certain type of ecclesiastical mind that sees words as a means of incantation rather than communication (except perhaps with the Almighty) and therefore resists the expression not merely of new ideas but of ideas in general.

More important still, the ideas of men like Millington three centuries ago and of Linnaeus two centuries ago were communicated only to a limited circle of educated people who understood Latin and who, being beneficiarier of the establishment, could be relied upon to support it even if they did not believe much of its creed. But a century ago, when Darwin published his great work, mass education had already begun and ordinary people were starting to be able to read books, or at least reviews of books, in the rapidly growing newspapers of the time; so there was concern that Darwin's ideas might fall into the hands of any Tom, Dick or Harry and cause him to question the authority of his masters and pastors.

But perhaps the chief reason why Darwin's work caused such a scandal – though it was after all only the logical conclusion of the findings of Millington and those who followed his lead – was simply that it appeared when it did. Though it did not make such a feature of sex as did Linnaeus' work, the whole idea of evolution and natural selection is based firmly on sex, because selection needs variation to work on, and variation comes from sexual reproduction. But in Darwin's time prudery had reached new heights – or depths. Members of the moral establishment were more easily shocked and outraged by the facts of life, particularly sexual life, than at any other period in English history. Only two years before *The Origin of Species* was printed the eminent gynecologist William Acton had published his book on

The Function & Disorders of the Reproductive Organs, which asserted with all the authority of his leading position that sexual activity weakened body and mind, and that the very suggestion that females were capable of sexual feeling was a "vile aspersion."

Darwin had observed several things about plant sex even before he published *The Origin of Species*, such as that flowers of the pea family produce very little seed if they are covered with net to prevent insects from getting at them. Later he published an account of the different sexual structure of the two types of primrose flower, one with the female and the other with the male parts exposed. He made many other detailed studies of such "heteromorphic" flowers: ones that, while hermaphrodite, have different floral structures, in one of which the female and in the other the male is prominent.

Darwin continued throughout his life to make careful observations of the interaction of flowers and their pollinating visitors. His books on the subject are classics, especially his *Fertilization of Orchids*, published in 1862, and his *Effects of Cross- and Self-Fertilization in the Vegetable Kingdom*, which appeared in 1876.

There is no doubt that Darwin influenced the study of plant sex more profoundly than anyone else in the nineteenth century. His observations on flowers and their pollinators led directly to the new science of "pollination ecology," in which the interactions between plants and the things that bring about their mating – insects, wind, water or whatever else is available – are studied.

After Darwin have come many other explorers of plant sex in its numerous forms, and many new findings were recorded by the Müllers, Hildebrand, Delpino, Trelease and Robertson at the turn of the century, culminating in the magnificent work by Knuth, completed by Loew, from 1898 to 1905.

This was based on *The Fertilization of Flowers by Insects*, by Hermann Müller, and set out to collect together, under the title *Handbook of Flower Pollination*, the knowledge that had been gained on the subject during the three centuries since Camerarius proved the existence of plant sex. The first volume of the English edition of Knuth's work, published in 1906, contains a bibliography of books and articles on flower pollination consisting of no fewer than three thousand seven hundred and forty-eight different items. One item missing from that mammoth bibliography was a publication by a forgotten monk in an obscure journal, which had appeared less than seven years after Darwin's *Origin of Species* and which was to have an even more profound influence on our knowledge of how sex works.

Born into a peasant family in 1822, Johann Gregor Mendel spent most of his life in the Augustinian monastery at Brünn (now Brno, in Czechoslovakia), where working in the garden he patiently established the basic facts of heredity. While the Darwinists and the anti-Darwinists fought loud battles over the theory of evolution, Mendel quietly set about finding out by careful experiments exactly how characteristics are passed on from one generation to another. The plants Mendel chose for his investigation were garden peas. He took pure strains with contrasting characters, crossed them with each other, and noted the progeny. Then he bred the offspring with each other and painstakingly recorded the results. In total he grew some thirteen thousand plants during the course of his experiments. The first generation offspring, he found, were all alike, and showed only one of the contrasting characters of the parents: this was therefore the *dominant* character. In the second generation, three-quarters of the progeny continued to show the dominant characters (round yellow and wrinkled green seeds), the first one, which had therefore not been lost but was *recessive*. He concluded that the hereditary factors we now call *genes* are separated in the sex cells and come together at fertilization; this is his "law of segregation."

When Mendel crossed peas with two pairs of contrasting characters (round yellow and wrinkled green seeds,), the first generation offspring all had round yellow seeds. The next generation gave the ratio nine round yellow, three round green, three wrinkled yellow and one wrinkled green. This led to the "law of independent assortment." Mendel's laws, now slightly modified, remain the basis for the whole modern science of genetics. Yet when he published the results of his experiments in the *Journal of the Brünn Scientific Society* in 1866 his findings were ignored by the establishment. Once again an amateur had dared to discover things the professionals had failed to notice.

Not till 1900, sixteen years after Mendel's death, were his findings rediscovered and their importance realized.

The modern period has seen students of animal behavior make important new contributions to the subject, and data on breeding systems and their pollinators are being collected and stored in computer banks. But no totally new and sensational discovery had been made during this century until the episode of the judge and the wasp with which this chapter opened. What new and unimagined sexual discoveries remain to be made is anybody's guess.

We have come a long way since the days of ignorance, superstition and partial knowledge, made worse by a prudish unwillingness to accept the overwhelming fact of sex as dominant throughout nature. We now know many things about the sex life of plants that might even cause such pioneer voyeurs as Millington to raise an eyebrow.

This book gives some of those discoveries about the sexuality of the plant world that have been established over the last three centuries. What those discoveries show is that there is no form of sexual expression which even the most fevered human imagination can dream up that plants have not already experienced and developed. Their sexual organs are of the most amazing diversity, ingenuity and complexity; their sexual practices are varied and often bizarre; their sexual appetites are prodigious. The female parts are mostly – though not always – passive, but the male ones can be most strenuously active, as in certain species where a trigger mechanism causes pollen to be ejaculated with considerable force. Who can say what plant sensations may accompany such emissions?

About the only thing that nobody has yet discovered in the plant kingdom is the female orgasm, or anything like it. Partly that is because there is no structure in plants similar to the nervous system of animals – though even that statement needs to be qualified, since on the face of it the sudden collapse of the leaves of the Sensitive Plant immediately when they are touched certainly gives the impression of a violent nervous reaction, which is why botanists call it *Mimosa pudica*, the word *pudica* meaning modest, or ashamed. And there are flowers that blush deeply after sexual encounters, as we shall see later in the book.

However, since there are no obvious nerves to transmit, or brain to receive, stimuli, no sensations in the way that we know them can be recorded. Besides, with no visible means of communicating their feelings to us it would be difficult for them to convey their experiences of arousal, ecstasy or just simple satisfaction, even if they have such sensations.

In any case, the likelihood – even if one were able to demonstrate and measure it, perhaps as some assert has been done already by means of electronic scanners or lie detectors – of anything resembling a female orgasm in plants is fairly remote. Signs of sexual arousal are there to be seen, as they are in animals: moistening of the stigmatic – that is to say the female – surface; heightened color, attractive scent, converging lines that draw admirers hypnotically towards the center of the flower and guide them to exactly the right spot. But such satisfaction (in the original Latin sense of "getting enough") as there is in the act of being fertilized seems to end there, with becoming pregnant; no discernible pleasure is involved. It is probable, though, that this applies not only to the plant kingdom but to all animals except the human species. Mankind alone seems to have developed the capacity not merely to be used by sex for the propagation of the race but to enjoy it for its own sake.

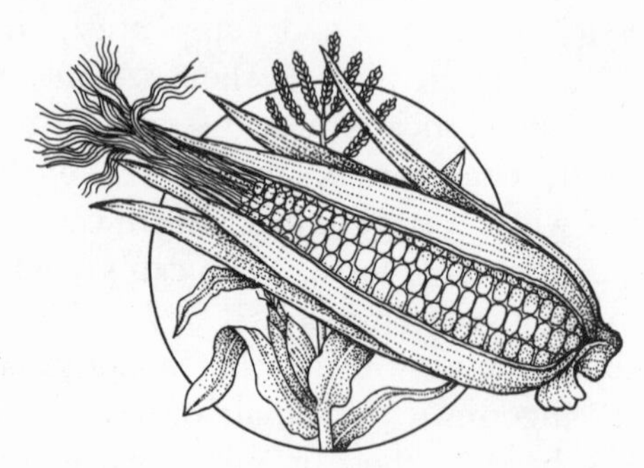

II

The Flowers and the Bees

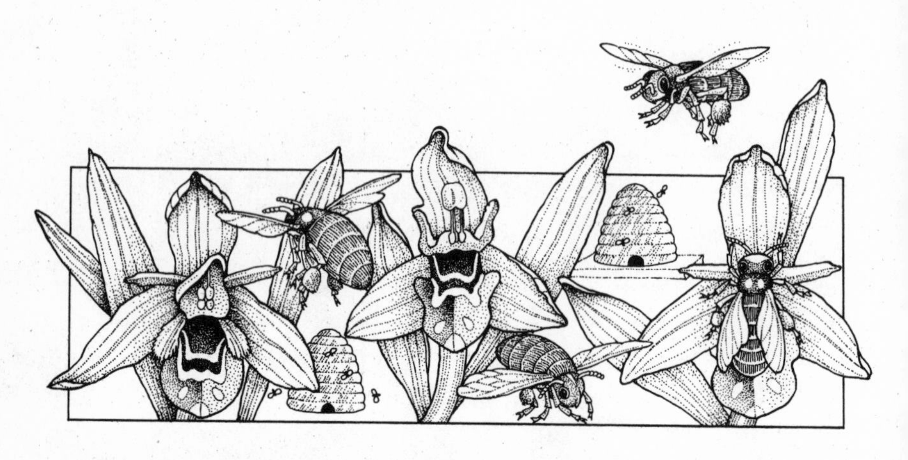

Before we go on to study the sexual structures of flowers in their remarkable variety and complexity, let us examine some of the ways in which they perform their function of providing for the production of future generations. In particular let us look at a few of the amazing number of ways in which they exploit whatever nature provides for their sexual needs: wind, rain, light, darkness, other plants – including the lowliest fungi – and animals of every size, shape and kind.

Some of the most spectacular sexual achievements of plants involve extraordinary feats of mimicry. Being for the most part, in comparison with the animals, passive and unable to move of their own accord, some of them, as we have briefly seen already, have managed to make themselves look remarkably like the insects and other creatures on which they rely for fertilization. By appearing to be a desirable mate for some agile creature with a powerful sex urge, they are able to use that sex urge for their own ends; and if the animal is fooled into a sexual performance with a plant instead of one of its own kind, that is just one of nature's many deceptions and of no concern to the plant.

Unlike human beings, who have their own taboos embodied in laws on the matter, the rest of creation makes no clear distinction between what is "natural" and what is "unnatural." Where sex is concerned, anything goes, and if the species can best achieve its purpose by sexual impersonation which amounts to a gross confidence trick upon a totally different species, it has no compunction or shame about that either.

As one of the most striking illustrations of the technique of sexual deception, let us take the case of the Bee Orchid, a close relative of the orchid which aroused a wasp's desire and a judge's interest, as recorded at the beginning of this book. Known to botanists as *Ophrys apifera* (which means "bearer of bees"), this charming species, which is spread throughout Europe, is to the human observer a visually attractive plant. A few people claim to be able to detect a faint sweet scent if they hold their nose close to the flowers, but most people find it quite without any perfume whatever; human beings are notoriously deficient in their sense of smell compared with the rest of the animal creation.

"Look, girls and boys," botany teachers say to the schoolchildren they take on nature rambles through the woods and fields to study wild flowers. "This is a Bee Orchid. Isn't it pretty? Don't you think the flowers look just like little bees?" And so they do.

But the human eye cannot easily distinguish between the sexes of bees, so to us the plump furry flowers look neither particularly male nor particularly female. To the more discerning eye of a real, virile made bee each appears to be a highly desirable female bee, offering herself seductively to him in a manner that says "Come and get me." An overpowering urge takes hold of the male bee. This is what he was born for. He zooms down onto the flower, mounts it, buries himself in its warm brown furry softness and attempts to mate with it. He starts to make a series of short, jerky movements and continues to do so for some time.

But something is wrong. However strenuous his lovemaking becomes, he does not achieve his sexual climax. Frustrated, he withdraws from the flower and flies off. During his exertions he has detached the pollen mass from the male organ of the flower, and this yellow, shiny mass clings to him by a sticky disk, called a viscidium, specially designed for the purpose. Thoroughly roused by now, and in a state of high frustration, he flies on to the next flower. There is nothing monogamous about his desires; all he seeks is sexual satisfaction.

He finds a new flower and goes through the same rough spasmodic courtship, during which the pollen mass comes into contact with the moist, sticky, concave female surface of the flower, called the stigma. The female surface, being stickier than the disk that holds the pollen to the bee's body, wrenches it away. And that is how cross-fertilization is brought about. Soon the ovary of the second flower begins to swell; it is, literally, "in pod." Most of the many thousands of embryo offspring that develop in the pod will die in early infancy; some will abort; but with luck a few will germinate and grow to continue the race of Bee Orchids.

The male bee continues from flower to flower, attempting to mate with each but never managing to achieve orgasm. To a human observer the whole process seems unnecessarily heartless: at least the flower might have given the bee his sexual satisfaction. However, that is not possible, for the simple reason that the flower lacks any structure like the sexual orifice of a female bee and so the organ of the male bee cannot be stimulated to the point of ejaculation.

Many observers, seeing frustrated male bees tired out after trying and failing time after time, have remarked that nature is quite ruthless and unsentimental; it is a matter of total indifference to the flower whether the bee is satisfied or not. If one goes more deeply into the matter, though, it becomes clear that the frustration of the bee is an essential part of the affair. It is positively to the advantage of the orchid that the bee should remain unsatisfied. A satisfied bee would relax after performing; having achieved his desire he would have no need to seek further amorous adventures. The urge would have gone. The bee would not be driven to copulate with other flowers, so pollination – the whole object of the exercise – would not be achieved. The Bee Orchid gains a distinct biological advantage by arousing and then frustrating its visiting suitor. It is a professional teaser, taking but never giving.

In fact as we shall see, there is another species of orchid in Australia with a floral structure that not only attracts male bees but can actually give them satisfaction; lucky bees have been observed on more than one occasion to reach the point of ejaculating sperm. What the flower can possibly gain from granting fulfillment in this way is hard to imagine; perhaps the species of bee involved is so easily discouraged that unless it gets satisfaction a fair number of times it will become so frustrated that it gives up trying.

Scientists call the process of false mating with a flower "pseudocopulation"; a nature photographer, with extraordinary patience and skill, has caught a bee in the act on a movie film that shows the whole performance, and can repeat it in a series of action replays as often as the audience wishes.

There is, though, a considerable hazard to the sex life, and hence to the future generations, of any species of plant that is quite as good an impersonator as the Bee Orchid. Since it cannot itself move in search of a mate, it has to wait for the mate to visit it. And that mate has to be a bee – and the right sort of bee at that – because the disguise is so convincing that the flower is not attractive to anything else. So what happens if when the flower has opened to display its sexual charms no bee comes along? Must it wait? And if so for how long?

The answer is sometimes a very long time. Indeed, some species of orchid which rely for fertilization on one particular type of insect, of perhaps a very rare kind, may have to remain open for weeks – or even months – before the right creature comes along. That is why many orchids are among the longest-lasting flowers. But a bloom that is going to remain open and sexually receptive for a long time may have to stand up to some very tough conditions – harsh, dehydrating sun, driving rain and blistering wind – without losing its attractiveness. So these long-lasting flowers have developed all kinds of devices to preserve their freshness and beauty against the worst that age can do. They cultivate smooth skins, with a cuticle that is hard, yet elastic enough to avoid bruising or chapping; they have tiny pores (*stomata,* the botanists call them) which can be closed tight in unfavorable conditions, because enlarged pores not only look unattractive but expose the delicate living tissues underneath the skin to the risk of drying out. They even cover themselves with a mask of wax to protect themselves against the ravages of time.

The Bee Orchid, however, does not go to such lengths to preserve its flowers. The result is that they may begin to fade within a few days. What happens if a bee does not turn up at all, or not until the flowers are so faded that he is no longer sexually attracted? The answer is that nature – or evolutionary adaptation if you prefer it – has taken care of that. In the last resort (and in spite of the often quite strict precautions that, as we shall see later, are built into the structure of many other flowers to prevent self-fertilization) a fail-safe device has been evolved by the Bee Orchid in case no amorous insect comes along.

While the flower is young and attractive, the central part called the column, which contains the genitals, is firm and erect. As it ages, the column shrinks till the pollen masses are ejected from the male organ. These, which contain the male reproductive cells, emerge on somewhat drooping stalks and hang down in front of the female part, which though no longer youthful is still sexually receptive. The slightest breeze which shakes the flower is enough to swing the limp stalks so that the male pollen is embedded in the sticky female recesses. In this case, it will be noticed, what human beings have come to think of as the normal order of things has been reversed. While the column was in a state of full erection, with the male organs on top, it remained impotent. Only when the most private male parts, the pollen masses, were exposed so that they drooped down to the level of the female organ did they become capable of sexual union.

So whether by means of a bee as its sexual partner or by its own devices the flower has been fertilized. If all goes well the pod will swell just as successfully and produce just as many offspring as if mating by a bee had taken place. In fact, though, as we shall see in the next chapter, such self-fertilization is not as good as what one might call the real thing because it tends to result in loss of vigor and adaptability. As Darwin showed as long ago as 1876, in his *Effects of Cross-Fertilization & Self Fertilization in the Vegetable Kingdom*, self-pollination can have a weakening effect. But at least the next generation has been secured, and in an emergency that is the main thing.

Since M. Pouyanne first described the strange process of pseudocopulation more than sixty years ago, many botanists have studied the ways in which flowers manage to arouse the sexual desires of the insects they deceive into attempting to make love to them.

The first sense to be acted upon seems to be that of smell. This confirms the findings of students of animal behavior, who have shown that scent is almost certainly the most primitive and powerful stimulus to the mating urge of all kinds of animals, from the lowest to the highest. Perfume manufacturers who promise sexual conquest to those who use their products are playing upon a very basic instinct. The scents of certain kinds of flowers have been thought to have such powerful aphrodisiac qualities that during some puritanical periods it is said that men and women were forbidden to walk together through a field of beans in full flower in case the strength of the perfume made it impossible for them to resist giving way to their sexual impulses on the spot.

Chemical analysis of the secretions that produce the enticing scent in these sexual deceivers among the orchids has shown that they closely resemble those produced by special glands in the abdomen of female bees and wasps. The perfume given off by these glandular secretions acts as a messenger to tell the males that there is a female in the vicinity waiting and hoping for their attentions. As has already been mentioned, the scent is not perceptible to most human beings, but some men can smell it quite distinctly. Unfortunately there is no scientific evidence at present on what sort of men have this specially keen nose for the mating scent of female insects; it would be interesting to find out whether they are also more responsive than their fellows to similar olfactory signals given off by females of their own species.

Experiments have shown that a male bee or wasp will recognize the mating scent of a female from quite a long distance away and be irresistibly drawn towards its source. Clearly this is a highly efficient arrangement on the part of the female, since it extends her range of allurement far beyond the point at which it would cease if she had to rely upon visual attraction alone. For the floral deceiver it is even more efficient; a female insect can, after all, move about in her search for a mate, but an orchid plant, however clever a mimic it is, must remain rooted to the spot.

The perfumes used by the orchid flower to lure the male insects are made up of a mixture of many different compounds with complicated chemical structures. Some carry a longer distance than the rest. These are the ones that first arouse the insect's interest and keep him alert and single-minded during his flight to discover the source of his excitement; they appear to have a directional function rather like that of the radar beam that enables an aircraft to home in on an airport long before the pilot can see the runway. As the male gets nearer, other scents get to work on his more immediate mating responses and stimulate him to a point of high sexual arousal.

As the flower comes into sight, visual stimuli play their part in arousing the bee's sexual eagerness still further. The shape and general appearance of the flower's most prominent part are just what the male's built-in instinct, or what some biologists would call "ancestral memory," cause him to expect of a female bee. The colors are exactly right too; indeed, they are more correct to him than they are to the human eye, because they contain the ultraviolet reflections that are perfectly visible to a bee but beyond the range of human vision.

To complete the visual deception, the lip and center of the flower are marked in precisely the right way, with the same pattern that in a real female bee guides the male towards her genital area. The directional signs evolved over vast ages to make sure that the male reaches exactly the right spot are faithfully reproduced by the flower; no forgery could be more convincing.

The last piece of mimicry that clinches the deception as far as the male bee is concerned is the texture of the flower. The furry surface of the lip strongly resembles the hairs on the genital region of a female bee and gives the male the necessary sensation which provides the final stimulus for him to make an attempt at sexual intercourse.

The bee's sense of touch must always be satisfied before he

can allow his mating urge to take over. It seems as if he needs the feeling of contact with a soft, warm, yielding body to reassure him that his sexual advances will be welcomed and not repulsed.

In short, males deceived by the Bee Orchid are the victims of a highly professional confidence trick which makes use of every one of their senses to mislead them.

Several other species of *Ophrys* in addition to the Bee Orchid practice sexual deceptions on the males of bees and wasps. Some are more convincing than others in their impersonation of females looking for a mate. All of them have more or less prominent, shiny protuberances, called pseudo-nectaries, which glisten just like the eyes of female bees or wasps, and so enhance the illusion. In addition, some go in for other optical tricks. The species with which this book opens, *Ophrys speculum*, has a lip like an oval mirror, of a glinting metallic blue color – hence its common name of Mirror Orchid – and is thickly fringed with rusty red hairs. On each side of the lip are short lobes which overlap the middle, looking remarkably like the folded wings of an insect at rest. At the front are dark red petals reduced so that they are no thicker than wires, and these closely resemble feelers or antennae. The whole appearance is similar to that of the female of a digger wasp called *Trielis* (or *Campsoscolia*) *ciliata*, somewhat larger than an ordinary honeybee.

It was through observing, over a period of many years, that only male wasps visited the flowers of the Mirror Orchid that M. Pouyanne first began to wonder. The flowers could not, he reasoned, be a source of nectar or food, or female wasps would visit them as eagerly as males. What could the flowers possibly be offering that was only of interest to males? He began to watch more closely. Every moment he could spare he spent observing and pondering, to try to find the answer to the mystery. Then one day quite by chance he noticed a male wasp having sexual intercourse with a female and realized that the position the wasp took up was the same as he had seen males adopt when visiting the flowers of the Mirror Orchid. The answer to the mystery had come to M. Pouyanne in a sudden flash. Up to that date the only suggestion from naturalists as to why the flowers of several *Ophrys* species looked so much like insects was that it must have been to frighten away marauding creatures which might eat or otherwise damage them. Rolfe, the well-respected expert on orchids, had recently mentioned the idea that the resemblance was a device to scare away browsing cattle. Now Pouyanne realized that the impersonation practiced by the flowers was not to repel but to attract. It was sex the wasps were after from the flowers. The process of pseudocopulation had at last been recognized.

Neither Pouyanne nor later observers have seen males of any other species of wasp than *Trielis ciliata* copulating with the flowers of *Ophrys speculum.* Biologists call such "faithful" flowers monophilous, which means literally "having only one lover"; there are several other members of the *Ophrys* genus that also attract only one kind of pollinator.

These "one lover" species of flower are not faithful through any moral scruples, but only because their mimicry is so exact and their physical makeup so specialized that not only do they appeal solely to one species of male insect but no other species would attempt to mate with them even if he wished. The very considerable gain to such faithful flowers is that they are so well adapted to the desires and the physique of their insect lover that they can expect an expert sexual performance from him, precisely suited to their needs. There is no danger of their being torn or otherwise hurt by the clumsy lovemaking of an unsuitable mate.

Among other "faithful" members of the *Ophrys* genus observed by Pouyanne in the course of his investigations was the species *Ophrys lutea*, a fairly common orchid in Algiers and around the Mediterranean coasts. The lip of this flower is a bright yellow, with metallic narrow blue "wings" on each side and a dark patch in the middle and at the base. It looks like a bee sitting on the flower. On many occasions Pouyanne watched small male bees of the species *Andrena senecionis* visiting these flowers. They made the same sort of jerky movements in attempting to copulate with the lip of the flower, but used a totally different position. Whereas the wasps he had first noticed trying to mate with the Mirror Orchid had placed themselves during the performance along the lip with their heads towards the center of the flower, so that when they emerged they had the pollinia sticking to their heads like horns, the bees attempting to mate with *Ophrys lutea* did so with their heads pointing towards the outside of the flower, so that the pollinia stuck to their abdomens instead. It seems that the lip of the flower must look to the males like a female bee lying head downwards on a yellow flower. The stigmatic surface of the flowers of *Ophrys lutea* is so placed that it is exactly right to receive the pollen from the belly of the bee instead of its back. The attempted mating by the bee does not look quite as comfortable as that by the wasp with the Mirror Orchid, because the bee has to turn round and back into the flower. But, perhaps the bee prefers things that way, and *Ophrys lutea* is merely catering to its liking for a different position. It is by no means unknown for insects – and other creatures as well – to prefer to copulate within some kind of enclosure, however slight; perhaps it gives them some sense of protection at a particularly defenceless and vulnerable time in their lives.

To investigate the reason for the bee's back-to-front position when mating with *Ophrys lutea*, Kullenberg tried an experiment. He cut the lip off one of the flowers, stuck it on the other way round, and waited. Soon a bee found the doctored flower and tried to have sexual intercourse with it – but this time it was the "normal" way round, with his head towards the center of the flower instead of his rear end. So it was clear that the bee was not reacting to the flower as a flower, but only to the position of what he took to be a female of his species.

Many other species of *Ophrys* appear to seduce wasps and bees as well, offering slightly different attractions according to the habits and preferences of their visitors. Anthony Huxley, in a chapter headed *A Floral "Kama Sutra,"* says that such plants can only be called prostitutes, but that seems unfair both to the flowers and to their insect "clients." A real prostitute will at least offer her patron genuine sexual intercourse, however perfunctory, with the possibility of a climax at the end, but the flower will offer no satisfaction at all, only frustration. On the other hand, while a prostitute makes her favors available to anyone who can pay the price, the *Ophrys* flower is interested only in starting a family; once it has been fertilized it gives up trying to attract any more visitors. It very quickly stops giving off perfume and its petals soon wither and fade. It is no longer in the business of seduction, only of pregnancy.

Clearly wasps and bees – at any rate male ones – are better than human beings at distinguishing between the different species of *Ophrys* which mimic insects. What is commonly called the Fly Orchid, *Ophrys insectifera* (or *muscifera*), does not look at all like a fly to the male of a wasp called *Gorytes* but like a female wasp of his own kind, which is why he has been observed in close embrace with it on several occasions. The attempt at copulation between this wasp and the Fly Orchid seems to be both the longest and the most strenuous of all. In their excellent book *The Pollination of Flowers*, Michael Proctor and Peter Yeo describe the performance in these words: "Often the wasp remains on the flower for many minutes, every now and then restlessly vibrating its wings and changing its position before settling down again and performing movements which look like an abnormally vigorous and prolonged attempt at copulation. While it is on the flower the wasp seems quite oblivious of the observer's presence." Perhaps the scent from the Fly Orchid has a hypnotic effect on the wasp; such is often the case with the perfumes given off by certain types of flower. However, in spite of the length and vigor of its lovemaking, *Gorytes* seems no more capable of reaching orgasm with the Fly Orchid than does his more quickly disillusioned relative with the Mirror Orchid. Nevertheless, the flowers of the Fly Orchid get fertilized, so from their point of view the encounter is well worth while.

Another *Ophrys* species that attracts insects to try to make love to it is *O. fuciflora*, known as the Late Spider Orchid because of its season of flowering and its rounded lip, which looks like the body of a plump spider. It is not spiders that attempt to mate with it, however, but large grey male bees of the species *Eucera tuberculata*, which Col. Godfery watched visiting it several times in the south of France; it was presumably to enable it to bear the weight of the heavy and rather clumsy body of this large bee that *Ophrys fuciflora* had developed its stronger and less delicate lip.

Many other examples of pseudocopulation, some proved and some subject to further observation, are given in the lavishly illustrated book *Orchid Flowers: Their Pollination and Evolution*, by van der Pijl and Dodson. They give one particularly interesting example which shows that other insects are involved in false mating with flowers besides the bees and wasps which for many years after Pouyanne's discovery were thought to be the only insects high enough in the evolutionary scale to respond to such a highly sophisticated form of sexual deceit.

The orchid in question grows in fairly high places, at between 2,500 and 3,000 meters, in Central America, from Colombia to northern Peru. It is called *Trichoceros antennifera* and it is found in rather dry parts among low scrub. There is an abundance of flies, feeding on the nectar produced by the very sweet-scented flowers of Climbing Hempweed, an evergreen member of the daisy family. Among the commonest species of flies in the region are those belonging to the *Larvaevoridae* (or *Tachinidae*), which spend their adult life sipping nectar and copulating, and their larval stage as parasites inside other insects, on whose bodies the female fly lays her eggs.

When the females are ready for mating they display themselves on a leaf or some other surface, and when a male fly passes near them they open and close their genital aperture, which glistens in the sunlight.

The lip and center part of the flowers of the orchid *Trichoceros antennifera* look so exactly like the females of one of the flies of this family, a species of *Paragymnomma*, that it has been said that they seem as if they might fly away at any moment. The reddish-brown and yellow markings in the middle of the flower and at the base of the lip, together with the side lobes that look just like wings, are quite enough to catch the attention of any passing male fly of the species. The rather stiff bristles on the flower are of the same color and texture as those on the abdomen of the female, and placed in just the same position, and glinting in the same way, as the genital orifice of an eagerly expectant female fly.

The male flies, excited by all the signals of sexual willingness, swoop down on the flowers. At closer quarters the resemblance to a female fly is much less convincing, and the male fly soon realizes his mistake; but by that time he has already struck the flower with enough force to cause the pollen mass to stick to his abdomen firmly, and be carried off by him. Just as easily fooled by the next flower he sees of *Trichoceros antennifera*, the male bee goes through the same performance again, and in so doing transfers the pollen to the stigmatic surface and so brings about fertilization.

So far all the instances of pseudocopulation we have dealt with have been shameless frauds. Whether wasp, bee or fly, the male has not achieved the satisfaction which all the sensory signals from the flower – by smell, sight and touch – entitled him to expect. The next example is different: the male not only seems to enjoy the experience but actually gets what he wanted.

The orchid in this case is an Australian genus called *Cryptostylis*, known as "tongue-orchid" from its shape. In a series of observations published, mainly in the *Victorian Naturalist*, over a period starting in 1927, Edith Coleman described discoveries made about the method of pollination of four species of the genus. The pollinating insect was an ichneumon wasp called *Lissopimpla semi-punctata*, which mated eagerly with all four species. Though there were considerable similarities both between the species of flower and between them and the female of the wasp, there were sufficient differences to oblige the male wasps to adopt slightly different methods according to which species they were trying to mate with. As described by Mrs. Coleman, the wasp, when attempting to copulate with the species of tongue-orchid *Cryptostylis leptochila*, had to back into the flower, arch its body and hang on tightly with his claspers. During the rather strenuous movements that followed, the pollinia became detached from the flower and cemented on to the wasp's back, in exactly the right place to make contact with the sticky female stigma of the next flower the wasp visits when he has finished with the first one.

With another species of tongue-orchid, *Cryptostylis subulata*, the male wasp had to make a different approach and use a different position. In this case the lip of the flower is considerably longer, hangs down more and is deeper, so that it wraps itself around the visitor. To the male wasp it must appear like a female stretched out fully and offering more of an embrace. After landing on the lip he backs for a considerable distance till he finds the right contour of the lip to clasp, by which time he has reached the middle of the flower and unwittingly detached the pollen and glued it to himself.

In the case of the two other species of tongue-orchids that the same wasp was seen to visit, *Cryptostylis erecta* and *C. ovata*, he was observed to use the same position and sexual technique as with *C. leptochila*, with slight variations to fit in with the small differences of floral shape.

The remarkable thing which seems to separate the mating of *Cryptostylis* by *Lissopimpla semipunctata* from all other instances of pseudocopulation so far recorded is that the wasp has been seen to reach orgasm and ejaculate seminal fluid. Indeed, he seems so delighted with the whole affair that Mrs. Coleman has reported that in many cases she observed he seemed to prefer the flowers to real female wasps.

So far we have only considered heterosexual pseudocopulation: attempts at sexual intercourse by a male with what he believes to be a female. Are there any examples of attempts at homosexual behavior between insects (or any other animals) and plants? So far there seem to be none. However, it is reported by van der Pijl and Dodson that when two male bees visit the same flower at the same time with sexual intent strange things have been known to happen. To quote their words, "The stimulated second visitor tries to copulate with the first male." That, however, can hardly be described as true homosexuality. What happens in such cases is that the second bee has become so aroused by the time he gets to the flower that he would probably try to copulate with anything of about the right size and shape; besides, the scent coming from the flower is a female scent and tells him that in spite of appearances the creature under him is a female. The mistake is soon realized and the two male bees go on their separate ways, to seek satisfaction elsewhere.

What might at first sight seem to be an instance of real, deliberate homosexuality has been observed in Ecuador. Certain species of the orchid genus *Oncidium* carry long sprays of small, bright flowers which on their slender stalks move in the slightest breeze like insects on the wing. Two such species, *Oncidium planilabre* and *O. hyphaematicum*, have sparkling yellow flowers with dark blotches and spots, which look, as they dance in the air, like the active and aggressive male bees of the genus *Centris*, of which several species abound in the region. On many occasions encounters have been witnessed between the real male bees and these flowers which resemble them. Such encounters are short, violent affairs, with the bees behaving very roughly towards the flowers. Though females of the same bees were seen to visit many other kinds of flowers, they seemed to have no interest in these two species of *Oncidium* and left them alone.

After the same thing had been observed many times, it was realized that the flowers did not represent sex partners to the male bees at all, but rivals. Male bees of the *Centris* genus are strongly possessive where their territorial rights are concerned. They each mark out a special territory (some observers suggest by smearing points at the boundary with scent produced by special glands) and guard it from intrusions by other flying insects, particularly males of their own kind. They perch themselves on a leaf or a shoot from which they can watch over their territory and at the slightest sign of an intruder they attack it. As soon as a breeze or other disturbance causes an *Oncidium* flower to dance in the air, the bee in whose territory the plant is tries to chase it away. If the disturbance continues, the bee will attack flower after flower, working himself into a frenzy.

At close quarters the resemblance of the flowers to bees is not very strong. But then it does not have to be. The flowers do not want the bees to make love to them; if they did, their disguise would have to stand up to close and intimate scrutiny, not only by sight and smell but by touch. All the flowers require from the bees is that from some distance away they should be mistaken for rivals for just long enough to cause them to be attacked. Fortunately they can rely on an extremely strong reaction from the bees in defense of their territory; by capturing males, marking them and then releasing them, naturalists have shown that their attachment to their home ground is so strong that they will return at once to mount guard over it.

The attack by a male bee on a flower that he sees as a rival is very quick and very precise. He flies straight at the flower and strikes it a sharp blow. In doing so he removes the pollen mass, which is stuck between his eyes by means of a quick-setting glue. Within a short time, the weight of the pollen has pulled down the stalk which holds it to the bee's head, so that it sticks straight out in front of him, ready to be driven exactly into the female part of the next flower he attacks.

By means of experiments using dead bees, it has been calculated that to pollinate a flower accurately the bee must strike it in precisely the right spot, with a margin of error of less than one millimeter. After many careful observations, it is reported that the bee rarely misses.

The term "pseudo-antagonism" has been coined for this mimicry by a flower of a rival to a pollinating male, so as to invite attack in order to achieve pregnancy. Another term has been proposed for a third kind of mimicry designed to bring about fertilization. In addition to pseudocopulation and pseudo-antagonism, it is suggested that there is a third type of impersonation by flowers to satisfy their sexual needs. For this third type of mimicry the term "pseudoparasitism" has been coined.

The behavior that gave rise to the idea of this third level of deceit was reported by F. Fordham in the *Victorian Naturalist* in 1946. The incident described in the report was yet again an example of violence displayed by an insect towards a flower. In this case, however, the attacker was reported to be a female. But the encounter was not a violent lesbian affair. The flower under attack was an Australian orchid of the genus *Calochilus*, meaning "beautiful lip." The species *Calochilus campestris* carries rather strange flowers of a greenish-brown color, with a very striking purple lip covered with masses of brown hairs. It was this lip which Fordham noticed being stung repeatedly by a wasp of the species *Campsomeris tasmaniensis*. Females of this wasp lay their eggs on insects, and when the grubs hatch out they eat the body of their unwilling host. On examination it was found that the lip of the flower looked like the sort of insects on which this parasitic wasp is known to prey. It seemed reasonable, therefore, to assume that what the female wasp was doing was to complete her sexual cycle by laying her eggs on the insect-mimicking lip of the flower, to provide her offspring with a good meal when they hatched out.

Unfortunately the sexes of wasps of this type are not very easy to tell apart, and it was not established beyond any doubt that the wasp molesting the *Calochilus* flower was really a female. If it was in fact stinging the lip it must have been a female, because the females are the ones with stings. It is, however, suggested by van der Pijl and Dodson that the movements of stinging and mating might have been confused, the vigorous up-and-down motion being much the same in each case. So perhaps the wasp observed by Fordham was really a male engaged in pseudocopulation. Certainly the furry lip and imitation eyes displayed by *Calochilus* give a fairly convincing impression of a female wasp.

So the case for pseudoparasitism of *Calochilus* by a wasp must in this instance be considered not to have been proved. However, as van der Pijl and Dodson say, pseudoparasitism is just as logical as pseudocopulation and pseudo-antagonism. Since, as we have seen, nature never seems to draw the line at using any and every type of behavior for the purposes of fertilization, it seems highly unlikely that such a powerful drive as the urge to lay eggs in the most suitable places to ensure a good life for the hatchlings will not be exploited by some other organism for its own benefit. There are somewhat similar instances on record. For instance, Darwin reported having found fly's eggs in the flowers of a species of *Cryptophoranthus*, the "Window-bearing Orchid," so called because it actually provides window-like slits at the sides of the flower for insects to enter.

There are many instances in the plant world of highly specialized adaptations by various species to the habits and urges of one particular kind of visitor on which they rely for fertilization. As we have seen, these very specialized adaptations bring considerable benefits. Monophilous, or "one-lover," flowers (in which we must now include "one-rival" flowers since the discovery of pseudo-antagonism) obtain quick, efficient satisfaction of their sexual requirements. But to every gain there is a possible loss. If there is only one kind of male able to satisfy them, and no specimen of that particular kind of male should happen to come along, their needs will remain unfulfilled. As we have seen in the case of the Bee Orchid, the only alternative is self-fertilization; but this goes against the whole purpose of sex, which is intended not for self-gratification but for the crossing of two different individuals. The self-pollination of the Bee

Orchid must be seen as a degenerate tendency caused by the lack of the right sort of bees. In some areas it has had to go over to self-fertilization completely, either because the right bees have disappeared or because the orchid has strayed into regions where there were no such bees to begin with.

Some related orchids overcome the problem by spreading their favors a little more widely. They are so made that they not only appeal to more than one kind of visitor but give more than one kind the opportunity to make a practical attempt at coupling with them. Botanists use the word "promiscuous" to describe flowers that are pollinated by more than one agent. What the flowers lose in the precision of the sexual act they gain by having more chances of being visited instead of having to wait for the one and only right male to come along.

A question which has puzzled many people is this: Why should a male insect actually choose a flower to mate with, rather than a female of his own kind? Except perhaps for the Australian wasp reported by Mrs. Coleman to prefer flowers to real female wasps, such males get no satisfaction, so why do they do it? The answer is that the male has no choice. To take a typical case, in those places where the Bee Orchid is fertilized by pseudocopulation, the flowers open early in the season, when no female bees are about. The first bees to emerge each year are all males. They fly around looking for a mate, but since there are no female bees to provide genuine competition they have to make do with the flowers. Later in the season, when real female bees appear, males who have survived their youthful frustrations transfer their attention from the flowers to the females. By then the Bee Orchid has finished flowering, which is just as well, otherwise there would be no proper mating between the bees, and so no new generation of unattached males ready and eager to be fooled the following year.

Many suggestions have been put forward by naturalists as to how this type of mimicry arose. Perhaps it is just the last stage in a series of smaller deceits in which visual resemblance to female insects was simply the finishing touch. There is, for example, an orchid called *Trigonidium obtusum* in Central America which hardly bears the slightest resemblance to an insect of any kind, and yet male bees of the genus *Trigona* have been seen trying to copulate with the flowers and so pollinating them. The part that excites them sexually is the ends of the petals, which have a glandular surface and presumably give off the right female scent, as well as feeling right to the touch. Several bees have been watched trying to copulate with the same flower at the same time, each taking a separate petal as its "mate." In the chapter on Mimicry and Deception in *Orchid Flowers: Their Pollination and Evolution* there is a drawing showing three bees hard at it. One would be tempted to call it a picture of group sex, except for the fact that the bees appear to be quite unaware of each other, each concentrating his entire attention on the petal he is grasping.

Both perfume and floral secretions seem to play a very important part in the sexual arousal of visiting insects. Sometimes the male visitor appears to be quite satisfied with just these two things, and does not feel the need to exert himself to express his sexual urge by attempting the movements of copulation. Mrs. Coleman recorded her observations of how the male bee *Halictus lanuginosus* pollinates the orchid *Diuris pedunculata* by forcing the lip of the flower back so as to get at the nectar ring at the base. The perfume is no doubt what arouses the bee and causes his reaction. Evidently the nectar combined with the scent acts as a powerful intoxicant; on many occasions bees have been seen with their mouthparts stuck inside the lip of the flower for hours on end, staying absolutely still except for small vibrations of the wings from time to time.

Without any visible attempt at copulation on the part of the bees, it may be said that the allurement practiced by this orchid is not truly sexual. However, only male bees were found to visit its flowers, never a female; so the attraction must be based on sex. In the words of Mrs. Coleman when reporting her observations, there was one powerful driving force behind the bees' behavior: "response to some mysterious attraction possessed by the orchid, partly scent, partly a resemblance to the female of their kind; but more probably to a marvellous imperceptible summons, which we humans can, as yet, only partly interpret."

No doubt many more cases of sexual mimicry will be discovered in the plant world. Most of those so far found have been among the orchids, but there is no reason to suppose that they are the only members of the vegetable kingdom capable of using trickery of this kind for their own ends. Already reports have been made of sexual intercourse between insects and the flowers of plants belonging to families quite different from that of the orchids. One such case, observed by Kullenberg, was of attempted mating by a digger wasp, *Tachysphex*, with the flowers of a tropical West African plant, *Guiera senegalensis*, belonging to a family related to the myrtles.

It now seems quite likely that even in the vast majority of cases of pollination by insects, where there is no suspicion of any attempt by the visitor to mate with the flower, the scent which attracts the insect appeals to some deep-seated sexual impulse. Even the workers from a colony of hive bees, who spend their days collecting nectar and pollen and are called sexless because they never copulate or produce young, may achieve a sort of sexual satisfaction from the odors and essences which surround them in the flowers. Chemical analysis has shown that many of the substances in floral perfumes are very like those which produce the body odors with which not only insects but other animals signal to each other that they are ready for sex. Among the most powerfully aphrodisiac of these floral perfumes, many people would put those of the large-flowered and heavily scented orchids first. That is one of the reasons why they stand in such high esteem as gifts on special occasions by men to their wives and girlfriends. It seems that the whole life style of a great many orchids, among humans as well as other animals, is that of the tender trap.

Sometimes the trap is not quite such a tender one, the flowers acting not so much like seducers as aggressors. There are two stiff whiskers in the center of the male flowers of many species of the orchid *Catasetum* (which means "downward-pointing bristle"). The female flowers are borne on separate spikes and often look quite unlike the male ones. The whiskers in the male flowers are very sensitive antennae, and are fixed to the pollen in such a way that they act as hair triggers. The moment a visiting bee touches the whiskers, the trigger mechanism releases the pollen with considerable violence and sticks it firmly on the bee, where it looks like a pair of horns. (The mechanism is so strange that after Charles Darwin had tried to explain it to the biologist T. H. Huxley he wrote in a letter: "I carefully described to Huxley the shooting out of the pollinia in *Catasetum*, and received for an answer, 'Do you really think I can believe all that?' ") The bee, which was lured to the male flower by a sweet, musky scent and was scratching at the lip to find where the odor came from when he touched the trigger that threw the pollen at him, will carry it around with him until he visits a female flower, where it will be ripped from him by the sticky stigma, leaving him half stunned from the encounter.

The Bucket Orchid, *Coryanthes macrantha*, makes use of imprisonment as well as enticement to get itself fertilized. The large fleshy lip of the flower is expanded into a bucket-like receptacle, above which are two knobs which produce a considerable quantity of water. This drips into the bucket for several hours before the flower opens, which it does by folding back the sepals and petals out of the way. By the time the flower is ready to entice visitors there is enough water in the bucket to give them a bath. A strong scent produced by the flower draws male bees of the genus *Eulaema* irresistibly to it. These males immediately go to a special patch at the base of the lip and scratch vigorously at it to get at the liquid from which the perfume arises. A few drops of the scented liquid are enough to make a bee drunk and incapable, so that he can no longer hang on and falls into the bucket. He lurches around in the water for a while, trying to climb the sides, but their slipperincess and his own drunken state combine to make it impossible for him. Just as it seems that he is about to drown, he finds a narrow tunnel at one side, through which he is just able to crawl. As he wriggles his way through he knocks off the pollen, which firmly attaches itself to him. But he is not out of the tunnel yet; a projection from its roof holds him down, and it may take half an hour of drunken struggling before he sobers up sufficiently to squeeze under the obstruction and escape. After he has recovered enough to be able to make his way to the next flower he will repeat the performance, only this time he will probably fall into the bucket of water even quicker, because of his wet wings, and in the course of his second visit he will deposit the pollen on the female part of the flower.

What possible purpose, it may be asked, can the Bucket Orchid have in keeping the struggling male bee so long imprisoned? The answer is quite simple, and the reason a very practical one. A mechanism triggered by the bee shuts off the production of scent completely, but it takes some time for all traces of perfume to vanish. If the bee could get out quickly there would still be sufficient scent to attract him back into the same flower; but after half an hour or more there is no smell left, so a bee that had spent that long trying to escape will no longer be interested in that flower and will go on to other ones with the scent he finds so attractive. A few hours later the first flower will be producing perfume once more. The timing device of delayed recovery of scent production makes it highly unlikely that the same bee with the same pollen still stuck to it will return to the same flower, and so self-fertilization will be avoided.

Many other orchids have similar devices for causing insects to lose their foothold and fall into a trap; some of these devices are so ingenious and efficient that the flower does not even need to bother to make the struggling visitor drunk first. But why should it be necessary for so much violence to be used in the fertilization of orchids, whether in attempting to mate or fighting to escape? The reason again is really very straightforward. The pollen of most flowers is fine, powdery stuff, which can be distributed by the lightest touch of an insect or other animal, or even by a puff of wind. The pollen of orchids is quite different. Instead of being powdery, it clings together in solid masses, called pollinia. These take a fair amount of force to dislodge them, especially when the male organ of the flower has a cover called the anther cap; this has to be prized off before the pollinia inside can be released, and considerable force is often needed to perform this act of circumcision.

Some sort of trap is necessary in many cases, as we have seen, to force visitors to make the extra effort which orchid flowers need in order to achieve sexual fulfillment. Many trap the visitors into physical bondage, as in the examples we have just examined, but the last word in sophisticated sexual ensnarement must be pseudocopulation. However, this is strictly a minority activity. It is not by such abnormal methods that most plants

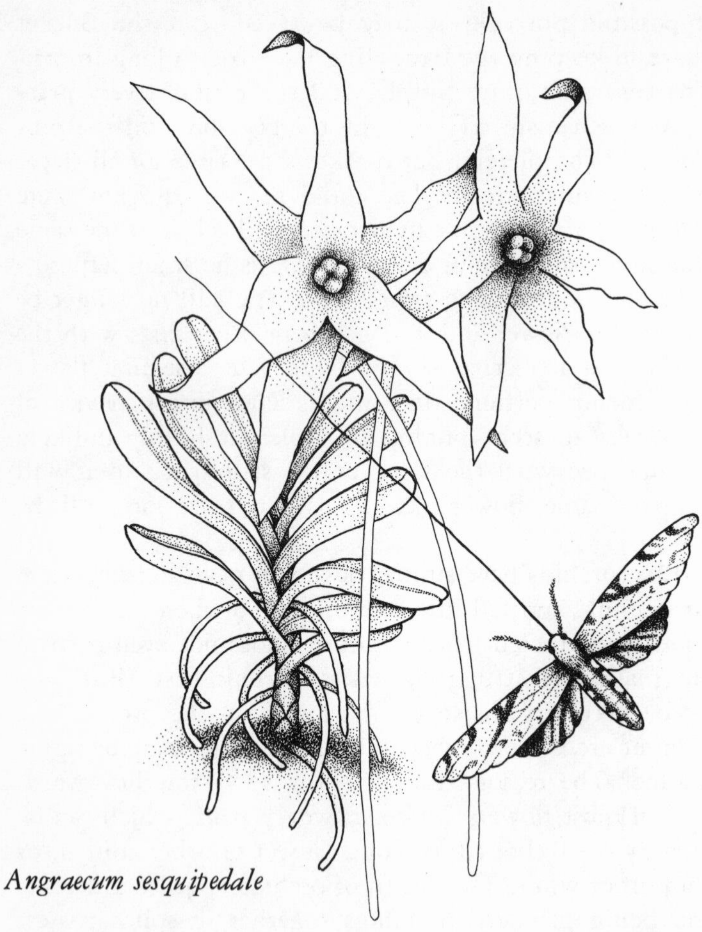

Angraecum sesquipedale

get fertilized. Normal sex among the flowering plants is not copulatory but oral sex.

To examine why, it is necessary to go back to the origins of sex, and see how the flowering plants developed alongside the insects and similar forms of life. This we will do in the next chapter. Before we do so, let us take a look at some examples of the remarkable ways in which flowers have developed their attractions to suit the mouthparts of the creatures that visit them and explore them orally.

An extreme example of adaptation to the truly extraordinary tongue of a pollinating insect is provided by the orchid *Angraecum sesquipedale*, from the hot lowlands of Madagascar. Its flower is one of the most spectacular in the whole orchid world, with heavy, waxy white petals five inches or more across. But it is the spur that is the most striking part of the flower and gives it the specific name sesquipedale, which actually means "a-foot-and-a-half"; this is perhaps something of an exaggeration, since the spur is in fact about a foot long. At the end of the spur nectar is formed; but what conceivable creature has a tongue long enough to get at it? Charles Darwin, after examining the flower, suggested that there must be an insect such as a large moth, with a wonderfully long tongue of between ten inches and a foot, which could probe deep enough into the spur to get at the nectar, and in doing so effect pollination. "This belief of mine has been ridiculed by entomologists," Darwin said. However, nearly forty years later his idea was found to be perfectly correct. A moth was discovered with a tongue of the required length, which was rolled up in a coil when not in use. It was named *Xanthopan morgani praedicta*, the last word in the name commemorating Darwin's prediction. Which came first, the long tongue of the moth or the long spur of the flower? It is often suggested that Darwin said that the moth's tongue evolved to suit the flower's long spur, but there is no evidence that he ever made such a statement. In view of the fact that flowers are the passive partners in the act of pollination by visiting creatures, it could be said that they have had to adapt themselves to the visitors, which have in a sense made the flowers the way they want them. If that is so, it could be argued that the flower had to produce its long spur to suit the moth's tongue. The answer is unlikely ever to be known; the most probable answer is that the moth's tongue and the flower's spur evolved at the same time.

The moth, as can be gathered from the fact that it took forty years to be discovered after Darwin's prediction, seems to be a rare and hesitant visitor. The flower will remain open and unfaded for weeks on end, its skin protected against drying out by its waxy covering. And, because its pollinating visitor is a night flyer, the flower is a luminous white to lure and guide the moth in the dark. (It is interesting to note that a similar discovery was made the other way around in South America. A moth was discovered with an enormously long proboscis, and many years later an orchid of the *Habenaria* genus was found with a spur of the same length.)

Angraecum sesquipedale represents one of the most advanced examples of adaptation to suit a discriminating visitor. There are many examples of plants which in order to satisfy their sexual needs cater to the appetities of much cruder visitors with much simpler mouths. One of the most remarkable is a plant that relies on exciting the interest of snails. It is described in considerable detail in a book from which we have already quoted, *The Love Life of Plants*, by the naturalist Raoul Heinrich Francé.

The idea that flowers could be fertilized by snails, he points out, is fairly unexpected for most people. As all gardeners know, snails are very destructive creatures. They do a great deal of damage by gnawing holes in the leaves and stems, and even the flowers, of many plants. To do this, they have been equipped with specially rough tongues, with which they rasp through the tissues of the plant.

The ordinary garden snail has more than fourteen thousand teeth; these are not set in the jaw like the teeth of an ordinary animal, but arranged on the tongue in rows. There are a hundred and thirty-five of these rows of teeth, with a hundred and five teeth in each row: tiny, hard, hook-like affairs which turn the tongue into a kind of saw or file. Such a coarse tongue may seem hardly a suitable instrument to apply to sexual organs, which usually require rather gentler treatment. Nature, however, seems to be able to make use of anything, no matter how unpromising, for the purpose of reproduction.

The pollination of plants by snails, though rare, is by no means unknown. Several accounts have been given of its occurrence, but usually it has been treated as a pure accident, in which a snail has quite by chance shifted a few grains of pollen from one flower to another. To survive a snail's attentions, a plant would have somewhat special requirements. First, it must remain undamaged, or not seriously damaged, when roughly licked by the snail's abrasive tongue. Second, the flowers must be in the right place at the right time. Snails usually spend the daytime in crevices or under dead leaves and other debris, and only come out at night. They do not like to stray very far from cover in their search for food. They hate having to move over dry, rough surfaces because they have to produce a good deal of extra slime to lubricate their passage across such places. They object to having to climb very high. So the ideal plant for them would be a low-growing one in damp surroundings near to suitable ground cover.

One such plant was studied with great care by Francé, until he was able to say he was absolutely sure that it had adapted itself to fertilization by snails. The name of the plant is *Alocasia*, and it is a member of the family called *Araceae*. Perhaps the best known member of the family is the plant known as the Arum-lily, which has a waxy white trumpet-shaped "flower" and is used a great deal in making up funeral wreaths. Because of its unspotted whiteness it has been held to represent purity and virginity; it is said to have been used in bridal bouquets in the last century, when virginity was a rather more highly valued commodity. It was also used in many Victorian flower arrangements for the home, but refined hostesses were recommended to cut out the central fleshy spike, which with its indelicately phallic look somewhat spoiled the virginal effect.

The *Alocasia* used to be grown a good deal in hothouses, but more for its handsome shiny leaves than for its flowers. The trumpet-shaped spathe opens widely at the top, then becomes narrower, and bulges out again towards the bottom. Inside this rounded chamber are the female flowers, hidden at the base of an organ called the *spadix*, which stands up stiff and erect, carrying the male flowers above the female.

The whole sexual apparatus is designed to cheat visitors, according to Francé, in a particularly artful way. The method it uses is called *dichogamy*, which literally means "two different marriages." It involves a complete change of sex. The female flowers open first, and remain open for some time while the male flowers above them stay tightly closed. Not till the female flowers have become old and withered do the male ones open in their turn. During the flowering period a smell of decaying matter is given off, rather repulsive to human nostrils but very attractive to creatures that live on and in refuse. Drawn by the smell, a snail may crawl into the spathe in search of the source of the delicious odor. If it does so during the second, all-male phase, it will become covered with powdery pollen. Not only did the snail Francé was watching find this pollen irritating, but the entrance to the withered female blossoms was closed to it. All a disappointed snail can do then, Francé commented, is to leave the inhospitable place as quickly as possible. But snails have rudimentary brains and do not learn readily from experience, so the visitor tries its luck again with the next plant. Perhaps this too is elderly and past the female stage, so that the snail only gets powdered over once more with pollen. Still having learned nothing, the snail may try again and again.

To quote from *The Love Life of Plants*, "If it is lucky enough to find a virgin, just developed into womanhood, it feels a glad surprise." The snail does not get peppered with pollen this time, and the tightly shut male flowers do not block the passage to the open female blossoms, from which a strong fragrance comes, promising the snail a great treat. It glides its slimy way through the narrow opening to the warm, inviting female place. However, what drives the snail on is the prospect not of sex but of food; being a hermaphrodite it has an unusual set of sexual urges in any case. Its single-minded ambition is to deflower the female blossoms in a thoroughly literal way, by gobbling them up. But as soon as its coarse tongue gets to work the female flowers defend themselves against the assault by a very effective anti-rape device.

The slightest wound causes a caustic juice to gush out, which scalds the sensitive skin of the snail and causes it acute discomfort. Even the walls of the chamber secrete the same burning liquid. In great pain the intruder makes its escape as quickly as possible, which with a slow-moving snail may not be quickly enough. If it does not get away in time it may be scalded to death.

During the time that the snail is crawling over the flowers with its body still powdered over with pollen from its previous attempts, it blunderingly succeeds in bringing about fertilization. And the more it writhes about in agony, the more thoroughly it will distribute the pollen and satisfy the needs of the female flowers. The snail receives no thanks for its trouble, however, only pain.

So long as the snail has not been too badly damaged in the process, it may repeat the same thing over again. Why, Francé asked, are only snails so dull? Might not an insect be fooled in the same way? The answer was that an insect's brain is so made that it would not be taken in by such a gross swindle; insects must be offered something: either sweet juice, or pollen, or flower honey; otherwise however attractive flowers may be they are ignored. But snails are more stupid. They are the greediest of all animals and gnaw at everything they meet; they rasp with their tooth-covered tongues, all night and half the day, any leaves that are not protected against them. That is why there are plants which lay traps for the snail's tongue by having tiny crystals, as pointed as needles, in their leaves; other plants protect themselves with hairs armed with barbs, acids, strong smells or poisons. So the poor snails wander hungrily from plant to plant, mostly in vain, and in the end they have to content themselves with decaying leaves in which the protective substances have lost their effectiveness. This, Francé concluded, really settles the question of why the *Alocasia* is avoided by insects but sought by snails.

Whether he was correct in asserting that only snails visit *Alocasia* is doubtful; its native habitats in tropical Asia abound in the kinds of flies which, as we shall see later, are constant visitors to similar trap-blossoms commonly used by members of the aroid family to deceive insects. However, there is no doubt that snails seem to have little or no learning ability. They are born stupid and remain stupid for the whole of their lives, if by stupidity is meant the incapacity to learn by experience. But nature is able to use anything for its purpose, including stupidity. If there were no stupid creatures around, plants like *Alocasia*, which rely on traps for the unwary, would not be able to get themselves cross-fertilized.

But even such dull creatures as snails may vary somewhat in their stupidity. None of them could be called exactly bright, but perhaps there are some that will not go on repeating the same pattern of behavior unless they receive some reward for their trouble. To cater to them, and at the same time to satisfy its own sexual needs, a plant of the lily family called *Rohdea japonica* has developed a remarkable method. It actually gratifies the greed of the visitors – which may include slugs as well as snails – by offering part of itself to be eaten. The flowers, half hidden among the leaves, give off a smell compared by some to bad bread, though Anthony Huxley, in his fascinating book *Plant and Planet*, goes further and describes the smell as that of bad breath. However, whether the odor is of bad bread or of bad breath, the slugs and snails find it extremely attractive.

The smell draws the visitors to the clusters of flowers, which are fleshy in texture, and on which the slugs and snails browse happily, eating away the juicy outer layers and leaving the remainder chewed and bedraggled. By gratifying their own desire for food, the visitors have gratified the plant's desire for sex. The missing portions of flesh from the flowers could be said to represent love bites, and the whole performance a thoroughly masochistic experience.

Evidently slugs and snails do not fancy the female organs, the stigmas, because it is reported that they make no attempt to eat them. However, in moving over stigmas, the visitors leave on them pollen which they have picked up from male organs during the course of their feasting. And that is how fertilization takes place. The process was described as long ago as 1895 by Loew, in *The Life of Plants*, and has been commented on many times since, but Faegri and van der Pijl say that the whole question of pollination by slugs and snails needs further investigation, to see how far it may happen by chance and how far some species of plant may have come to rely on it.

It is sad to have to record that if it does rely on this peculiar method of pollination, the *Rohdea* may be in danger of not being able to propagate itself sexually for very much longer. Closely related to that old favorite houseplant the Aspidistra and almost as easy to cultivate, *Rohdea japonica* has been extensively collected from its native habitats in Japan and parts of China, in order to be grown in gardens and greenhouses by enthusiasts. As a result of this and of clearing and building, it has disappeared from many places where it used to grow wild.

The plant is called Omoto in Japan, where it has long been a favorite among rich fanciers, who have been known to pay very large sums of money for specially desirable specimens. According to the American *Standard Cyclopedia of Horticulture*, "Retired persons of means often spend their declining years in the culture of this interesting plant, on which a number of very beautiful books have been written." Unfortunately, such dedicated fanciers would make it a point of honor and pride never to allow a snail or slug among their plants, so if the species is reliant on these mollusks for pollination its sex life seems to be doomed; but then so is the sex life of many other plants taken from the wild and kept in captivity.

In this chapter we have seen many examples of deceit practiced by plants on visitors. All these cases of deceit are sexual in origin – as indeed some philosophers and theologians have suggested is the case in a wider context – because they all stem from the need of the plant to get fertilized, for which purpose it will use any trick, however unscrupulous it may seem to a human observer.

Let us end the chapter with what is perhaps the most unscrupulous trick of all. The plant concerned is *Nymphaea citrina,* a water lily with starry yellow flowers which grows in central Africa. The flowers go through two distinct stages, involving a complete change of sex. On the first day the flower opens it is in the female stage; the male organs, the anthers, are closed and point straight upwards, forming a slippery surface. The flowers are promiscuous – that is they attract many different visitors – and an assortment of bees and other insects land on the surface formed by the closed stamens and slide down into the bowl-shaped female part beneath, which contains a thin, sugary liquid in which the visitors are quickly drowned. Some of the bigger and more agile ones may manage to get out, but not many.

In its second stage the flower becomes visibly male; the stamens converge to form a sort of dome-shaped hood over the female bowl, which is hidden from view. As the stamens ripen they turn outwards and form with the inner petals a sort of stage on which visiting insects can crawl about. In doing so they pick up pollen, which is carried by them to the next flower, which if it is in the female stage will throw them into the bowl of liquid to drown. But in drowning they will shed the pollen and so, after the bowl has dried out, which it will quickly do during the male stage, the flower will be fertilized.

Since for the visitors it is a one-way journey, with no chance of returning, the flower must rely for fertilization on the chance that the visitors have been on a male flower before drowning. However effective the flower's technique may be in the short term, it must be rather questionable in the longer term; killing off its pollinators must put future fertilization at risk.

The same applies to all forms of deceit, which means promising something – usually sex or food – and then not delivering the goods. In evolutionary terms, it has proved better all around for plants to give as well as to receive.

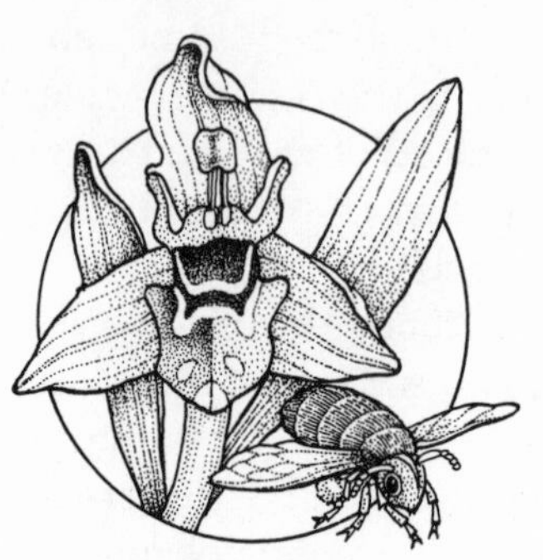

III

Why Sex?

In 1664, just twelve years before Sir Thomas Millington's discovery of plant sex, another Thomas was responsible, as star prosecution witness, for the condemnation of two unfortunate women, Amy Duny and Rose Cullender, at their trial for witchcraft.

Thomas Browne, the man on whose evidence the women were found guilty, was not only a successful physician with a fashionable practice in Norwich but a well known author, now greatly respected for his contribution to English literature. A few years after the trial he was honored by the king with a knighthood for his services. Sir Thomas Browne's writings, which include *Christian Morals* and many similar tracts, reveal him as a man who felt the deepest disgust at the sexual act performed by human beings and other animals, which he compared unfavorably with what he believed to be the innocent sexlessness of plants.

The evidence of such an eminent and respectable man was quite enough to convict the two women, who were believed – like other victims of witchcraft trials – to have had sexual intercourse with the devil. Understanding something of the psychopathology of witch-hunters, as we now do from studies of diseased minds and from such enlightening books as G. Rattray Taylor's *Sex in History*, we realize that the persecution of witches gratified their unconscious erotic desires, turned to sadism by an obsession with, and at the same time fear of, sex.

Sir Thomas Browne had already, twenty years before the Norwich witchcraft trial, shown signs of personality disorder about sex. In his book *Religio Medici* he had expressed both his own sexual hostility and the gathering puritanism of the age in these words: "I could be content that we might procreate like trees, without conjunction, or that there were any way to perpetuate the World without this trivial and vulgar way of coition: it is the foolishest act a wise man commits in all his life; nor is there any thing that will more deject his cool'd imagination, when he shall consider what an odd and unworthy piece of folly he hath committed."

That oft-quoted passage is one of the clearest revelations in literature of the feelings of guilt and shame that the subject of sex has produced in so many anxious people over the centuries. Put into present-day language it could be found among the letters in any modern magazine dealing with sex problems; what might be termed the Browne attitude is still a fruitful cause of the impotence, frigidity and other difficulties that cause sufferers to seek help from sex therapists such as Masters and Johnson.

Patients are on the way to the recovery of sexual health when they can understand and accept the sexuality of the whole living world, vegetable as well as animal. Sir Thomas Browne, however, seems to have remained ignorant of plant sex to the last. He died on his seventy-seventh birthday in 1682, the very year in which Grew published the text of his Royal Society lecture putting forward the bold new idea of sexual activity among the flowers. It is most unlikely that Browne ever learned of the idea; so he was able to die in peace, with his illusions about the sexless nature of plants unbroken.

Why did sex ever come into the world, and why has it aroused so much antagonism, at any rate from certain members of the human species?

The answer to both questions is the same. Sex means variety. That is what it is all about: to ensure that there are always differences between individuals, so that no two sexually produced living things are ever exactly the same as each other.

Without variety there could never have been evolution. The survival of the fittest demands that there should be differences between individuals, because if all were identical there would be no fittest to survive, or for that matter unfittest to perish.

Sex, in short, is not like a mass-production factory, turning out identical articles to a standard pattern. It is more like a creative artist, producing something new every time: occasionally a masterpiece of lasting value, sometimes an unsuccessful attempt only fit for the scrap heap. But by introducing variety, sex also introduces uncertainty. That is probably the basic reason why people with authoritarian tendencies have such a deep fear of sex. Insecure people, unsure of their own capabilities and finding difficulty in adapting to change, seek security and reassurance in rigid conformity, dislike the unusual or unexpected, and govern their lives by a fixed code of behavior, based on unquestioning acceptance of a set of rules. They prefer faith to enquiry and revelation to experiment. They are conservative, respect rank, ritual and uniform, and resist change.

Their counterparts in the vegetable world are those plants that reproduce themselves without sex. They are all alike, and they can be very successful so long as conditions remain exactly the same. Should those conditions change, however, the species may collapse, and even disappear completely, because lacking the necessary amount of genetic variety, it will not produce the odd, different individual, which through its very oddity may be better suited to the new circumstances. It may be taller or shorter than the rest, thicker or thinner, hairier or smoother, able to withstand more cold or heat, drought or flood, light or darkness, and so be better equipped to survive the changed conditions.

Sex was one of the two inventions of nature on which evolution has been based. The other one was death.

Death has been just as important as sex in the development of more advanced forms of life, and just as much hated and feared by rigid minds. Most people resist change; death clearly represents change; therefore they resist the very thought of death. The notion of eternal life was invented so that they can continue unchanged through eternity. It is small wonder, then, that Sir Thomas Browne, who felt his dignity so threatened by the idea of sex, should feel the same way about death. "I am," he wrote, "not so much afraid of death as ashamed thereof; 'tis the very disgrace and ignominy of our natures."

To examine how the twin mechanisms of sex and death came into being, we must go back to the beginnings of life on this earth. Present opinion among most scientists is that the first living thing appeared in the oceans somewhere between three and four billion years ago. It was a chemical molecule or combination of molecules similar to the others which surrounded it in that primeval soup. But there was a vital difference: it was able to reproduce itself. It may have been crudely simple in comparison with the higher forms of life which have appeared since, but it was alive.

It was quite without sex, and it never experienced death as we know it. It was formed of a single cell, and it reproduced simply by splitting in two. The two new cells could not be said to be children of the first; they were just parts of it, split from each other when it reached a certain size and broke in half. No doubt the same thing had happened many times before, when compound substances formed by the coming together of groups of atoms were broken apart again into simpler pieces by the violent physical forces then at work. The difference this time was that instead of being smashed into simpler components each separate half remained a duplicate of the original. Reproduction had begun.

Those earliest living organisms could not be called plants in the sense in which that word is generally understood. They lacked one essential ingredient. They were more like what we know as viruses: standing at the borderline between inert chemicals and living things, they obtained their food from the salts in the water around them. Modern viruses are different; they feed themselves from the living organisms whose bodies they invade and in whose tissues they multiply, causing diseases such as influenza, polio and measles in human beings, foot-

and-mouth disease in cattle, and tremendous losses among plants of all kinds, from food crops like potatoes, tomatoes and cereals to flowering plants grown for ornament. It is thought that many of these modern disease-causing viruses may be fallen forms of life which have degenerated from somewhat higher organisms and become parasites.

The thing all these earliest life forms lacked, and without which they could not become plants, was chlorophyll. Until the appearance of this substance none of those primitive organisms could be said to have been self-supporting; all of them were in a sense parasitic, because they could only grow by feeding upon the substances already existing in the water surrounding them and the gases above it. They were rather like modern mankind, exploiting the limited and diminishing resources of existing fossil fuel for its immediate needs but doing little or nothing to make use of outside sources of energy so as to add to the world's store of essential supplies instead of burning it up.

From an examination of the fossil record preserved in ancient rocks, and by the use of advanced chemical analysis, and the latest microscope techniques, scientists have calculated that there were already organisms containing chlorophyll some three billion years ago, only a few hundred million years after the first forms of life appeared. Because of this chlorophyll – the green pigment that gives the leaves of plants their color – these organisms were able to make use of solar energy, a source of power that mankind is only now, some three billion years later, trying to harness before the store of fossil fuels runs out. The process by which chlorophyll, powered by the sun's rays, is able to manufacture sugar and other carbohydrates from water and carbon dioxide is known as photosynthesis, which means literally "putting together by light." In its complex molecular structure, which forms an efficient chemical processing works, chlorophyll closely resembles the pigment in the red blood cells of human beings and other vertebrate animals (and some invertebrate ones such as earthworms) called hemoglobin, which is as vital to the life processes of animals as chlorophyll is to those of plants. And just as there are several blood groups, and different kinds of hemoglobin for different animal species, so there are many different kinds of chlorophyll, with slight variations in chemical structure.

The various chlorophylls are found in different combinations in different forms of plant life, but there is one kind called chlorophyll A which exists in every plant. In the blue-green algae, chlorophyll A is the only one; and since blue-green algae were no doubt the first plants, it seems that the A type is the original chlorophyll, three billion years old. (There exist a few photosynthetic bacteria, on the borderline between animals and plants, which contain substances similar to plant chlorophyll called *bacteriochlorophylls*; such bacteria can make use of sunlight to power their life processes, but they can also find all their requirements from the chemical substances around them.) It is thought that algae and bacteria came from a common ancestor, but that while the algae became industrious workers, building themselves up by manufacturing more elaborate materials from simpler ones by means of photosynthesis, bacteria went the opposite way, by taking existing foodstuffs that happened to be around and getting energy by digesting them, breaking them down from more complex to simpler substances in the process. Since foodstuff has to come from somewhere, bacteria became either scavengers, feeding on nonliving materials, or thieves, invading living organisms and feeding on them. It could be said that bacteria took the lazy way, since breaking things down is easier than building them up. However, bacteria have played, and continue to play, a vital and indispensable part in the world of living things. Without bacteria, the dead bodies and excreta of animals and the dead remains of plants would be piled up everywhere, since there would be no bacterial action to break them down again into simple chemical substances which return to the earth to restore fertility and to provide the food that enables other plants to grow. Without bacteria, certain crops such as beans and peas would not be able to take from the air the nitrogen, vital to growth, which at present is extracted by colonies of nitrifying bacteria that inhabit nodules on the roots of such plants.

True, without bacteria we would not have the diseases caused by parasitic kinds, such as pneumonia, diphtheria, tuberculosis and typhoid; but we would also be denied some of the most powerful antibiotics to fight disease, such as streptomycin, the natural source of which is the bacterium *Streptomyces griseus*.

So, long before sex was invented by nature, the bacteria took a different turning from the plants and went their separate way, making their own contribution, for good or ill, to the world we know today. And they have remained, to all appearances, quite without sex right up to the present. They multiply simply by splitting in half, or by other forms of nonsexual reproduction through spores, either passively carried by the wind or self-propelled by means of whip-like tails called flagella. Sir Thomas Browne would presumably have approved of them, because they reproduce in the manner that he wrongly attributed to trees, without that sexual "conjunction" he found so distasteful. Or so it would have seemed until recently; but scientists have now demonstrated that some forms of bacteria reproduce by a kind of sexual intercourse, so even the bacterial world has developed tendencies that would have caused Sir Thomas distress and disillusionment.

Since our subject is the sex life of plants, we must leave bacteria and return to the blue-green algae which by the invention of chlorophyll brought the plant world into being. Like all prototypes, whether of living beings or of modern household equipment, they lacked the refinements and complications which were introduced into later models, increasing the number of working parts and so the chances of breakdown, and consequently needing a repair service as parts wore out or ceased to function properly. And that, it could be said, is how sex came along, as part of the repair service. It was not till later that sex took on its other role of originating new models, and death had to be introduced to get rid of obsolete ones.

The span of time between the appearance of the first living things and the development of sex is thought to have been well over two billion years. During that long period, forms of life of ever-increasing complexity came into being, but it was not until that complexity had reached a certain stage that plants were ready for sex. The most primitive organisms were, in a literal sense, too crude for it.

As we have seen, one of the most important stages in the emergence of forms of life high enough on the scale to engage in sex was the invention of chlorophyll. That made possible all the development of higher organisms: not only the green plants themselves, but the fungi, and later on the animals that began to appear, feeding directly or indirectly on the materials manufactured by those plants.

The first plants, the blue-green algae, known together with bacteria as *procaryotic*, differed from all later forms of life in having the genetic material by which they reproduce themselves loose in the cell in the form of simple threads of DNA (*deoxyribonucleic acid*, the basic stuff of nearly all living things), by means of which they are able to go on making copies of themselves in their descendants. With such a simple cell structure, no sexual intercourse would seem to be possible in any ordinarily accepted sense of the words, since there is no recognizable differentiation into male and female.

However, the earliest form of coupling is thought to have occurred among the simplest algae. It could hardly be called true sexual union, but it was union of a sort; perhaps it is best described as pre-sexual union, since it appears to have been a necessary stage in the development of true sex.

What may have happened is this. These primitive one-celled organisms, as we have seen, multiplied by the simple splitting of one cell into two, four, eight, sixteen and so on. At some point in time, the reverse is thought to have occurred; instead of one becoming two, two became one. Perhaps the two cells concerned had been damaged in some way, so that their walls were ruptured; their contents mingled through broken skin and they became united into a single cell. It has been suggested that such fusion may have happened for nutritional reasons; following repeated splitting of an organism several times in succession, resulting cells with depleted reserves may have recombined to pool their resources. Perhaps the coming together happened simply because two cells happened to collide; it would seem appropriate if the process that was eventually to lead to sex had started by two individuals banging into each other.

This primitive coupling, as already pointed out, was simply a repair job. In fact it is probably wrong to call it coupling at all, since a couple would be thought of by most people as two different individuals. These two were not. They were parts of the same organism, identical in composition, which after being separated were joined together again; to use the word coupling to describe their union would be as inappropriate as to use it to describe the sewing back of a finger tip that had accidentally been cut off. Even the term self-preservation seems unsuitable, since that implies the danger of death; but there was no fear of that, because the destruction of both the damaged cells would have had no effect on the existence of the organism, consisting as it did of untold millions of separate pieces in the form of absolutely identical cells.

Such an organism, dating back to a time long before sex was invented, and with it death, can be said to be immortal; the simplest single-celled blue-green algae which live in the oceans to this day are not merely descendants but actual parts of that same organism, and so could be described as at least two billion years old. The organism could only be said to be dead if all those separate cells were destroyed together. And that seems hardly likely, since they can put up with extremely tough conditions; they will even survive temperatures of up to 75°C (nearly 170°F), which would kill any less primitive plant. This is because of the simple nature of the cell and its contents, which no doubt had to face similar extreme conditions when the first of their kind emerged. But the same simplicity of structure also made it impossible for them to engage in higher activities like sex, which demands considerably more complicated natures from its practitioners.

For the next several million years after those two cells merged with each other, the same thing must, it is surmised, have happened time and time again: casual and accidental pairing of identical cells, with no change and no variation. Since there was no such thing as sex to introduce variety, the only way in which any change could be brought about was by mutation. This was an entirely haphazard affair, relying on a chance event such as exposure to radiation, chemical contamination, mechanical damage or malfunctioning of the cell to change part of the genetic material. Usually this happens to a single gene, one of the units of inheritance, and causes it to alter, so that it changes the character of the organism in some way.

As chance alterations to genes are the result of accidents, most mutations are for the worse. They either kill the organism outright or so cripple it that it is placed at a disadvantage. However, in rare cases the mutation is not destructive and may even bring about a change for the better. It will then be carried on into the next generation and may, if it still has survival value, lead to a new strain different from, and possibly even superior to, the old one.

One day, let us assume (and this is conjecture, because the fossil record does not give us the details), two primitive cells of blue-green algae came together as had happened many times before; only on this occasion the two cells were not quite the same, because one of them had undergone mutation, so it was slightly different. The difference was not big enough for one cell to reject the other as a foreign body, but it was big enough to prevent them from merging completely; the contents of the mutant cell, instead of mingling with the contents of the other so that they became indistinguishable, stayed together and formed a separate pocket inside the other.

It may not have happened exactly like that, but in some way a new type of cell began to appear, with a nucleus separated from the rest of the cell's contents – the *cytoplasm* – by its own nuclear membrane. Such a cell, called *eucaryotic*, was a great advance on the simple structure which had gone before, and it has formed the basic unit of every living thing since, plant or animal. The organization of the cellular material into distinct parts introduced the whole notion of difference of function, which was to lead to the specialization not only of different parts of the same cell but later of different cells. It was also to lead to sex, based as it is on difference.

As the first introduction of variety was the work not of sex but of mutation, certain modern scientists believe, as we are reminded by Anthony Huxley in *Plant and Planet*, "that sexual reproduction is in fact unnecessary and that evolution could have taken place by mutation alone." The belief suggests that perhaps certain scientists may have had unsatisfactory sex lives themselves and consequently developed the Sir Thomas Browne syndrome and wished that sex had never been invented. The speculation was, in any case, a useless one; whether Browne, or some scientists, or anyone else liked it or not, sex was in fact invented. It had probably become inevitable as soon as living things had reached a certain stage, first of differences within themselves and then of differences between themselves.

The outcome of those differences has been the development, over the billions of years since life began on this earth, of a fantastically rich and varied plant world. Once the production of new forms had started there was no stopping it. To quote Anthony Huxley again, "One is tempted to postulate a dynamic natural force – what Henri Bergson called *l'élan vital* – which cannot stop itself from infinitely producing novelties – a kind of cosmic doodling, an innate urge to change and modify."

As a result of this "cosmic doodling" there are estimated to be something approaching four hundred and fifty thousand species of plants living in the world today. The majority of these species belong to the *Spermatophytes*, or seed plants, which dominate our present landscape. These in turn are divided into two groups: the *Gymnosperms* ("naked seeds"), which were already flourishing two hundred million years ago and are still represented by the conifers and their kind, and the *Angiosperms* ("enclosed seeds"), going back only a hundred million years or so and comprising the most successful development in the vegetable kingdom, the flowering plants, with something like a quarter of a million different species.

The normal way in which advanced forms of plant life – and indeed all other forms of advanced life – reproduce themselves is by means of sex, though some degenerate types have given it up.

Because they were the latest to evolve, the seed plants are often referred to as the "higher" plants. However, the word "higher" is avoided by many scientists because it suggests that the lowlier plants, which are inferior to them in a strictly factual sense – that is, they mostly tend to grow close to the ground or stay in the water instead of growing upwards and brandishing spectacular organs such as flowers – are also in some way inferior to them in value. As Freud pointed out many years ago, since people started walking upright and so raised their organs of sight and smell above the ground, they have been inclined to devalue things lower down and to find such things "beneath them" in more than a physical sense. Perhaps that is why they have so persistently denigrated sex as part of their "lower" nature, since their sexual organs are down below their heads most of the time.

There is nothing less "important" about the more ancient types of plant than about the newest and gaudiest flowering species. All have their place in the scheme of things. True, the kinds that existed before the flowering plants came into being have been at it longer, but they are none the worse for that. Some of them have shown amazing survival value, having remained unchanged or little changed for vast periods of time, during which many experimental flowering species have come and gone, leaving nothing but a few fossil remains. Besides, these ancient types of plant have, in addition to the ones that remain more or less unchanged since earliest times, evolved along their own lines into a bewildering variety of forms and shapes, with a remarkable assortment of methods of reproducing themselves, both sexual and nonsexual. Since it was among plants of this type that the first fumbling attempts at sex were made, without which the more polished techniques of the "higher" plants could never have been developed, we will examine them first.

Because to the naked eye plants of this early type show no sign of having sexual organs, they are often called *Cryptogams*, from the Greek for "hidden marriage." That was the name given to them by Linnaeus to distinguish them from those with sexual organs clearly in evidence, still sometimes referred to as the *Phanerogams* (from "visible marriage"), which include all the flowering plants and are now usually called the *Spermatophytes*.

For some time after sex had been accepted as an inescapable fact of life among the flowering plants, many people believed that there was no sex at all among the cryptogams; one group of plants, they said, even though lowly, had escaped the sexual urge which had seized the newer and flashier members of the vegetable kingdom. They were of course wrong, as always. Sex seems to crop up everywhere, in all groups of loving things. Some of the simplest and most primitive organisms do, it is true, appear to manage without sex in any form, showing no signs of sexual organs or sexual behavior. However, during the past few years so many assumptions have been proved false, so much unsuspected sexual activity has been uncovered by the use of new scientific methods, that much of what has been accepted as sexless reproduction may prove not to be. As Professors Faegri and van der Pijl say in *Principles of Pollination Ecology*, "Modern biology has made spectacular advances during the last few decennia. The results of genetical, biochemical, and other investigations have given us a new insight into phenomena of life never before understood, hardly ever perceived."

When people call sexual attraction between themselves a matter of chemistry, they are saying exactly what modern investigators are finding about the sexual reactions of plants to each other. Some of the most remarkable and exciting discoveries are being made among the cryptogams, whose "hidden marriages" are pried into in the most intimate detail by investigators using the very latest laboratory equipment. Let us take a look at a few of the experimental findings, which give us a glimpse of the very origins of sex.

The group called by Linnaeus cryptogams has since his days been split into three separate divisions of the vegetable kingdom. The first of these divisions, the *Thallophyta*, contains the most primitive plants, including the algae and the fungi, and is recognized by the simple plant body, called a *thallus*, which is not differentiated into root, stem and leaves. The second division is the *Bryophyta*, containing the mosses and the liverworts. The third division is the *Pteridophyta*, consisting of the ferns and their kind. We will examine plants from each of these three divisions to see how sex has developed from simpler to more complex forms.

As we have seen, sex is based on difference. And the most basic sexual difference, as we see it in the higher plants and the animals, is between male and female. In some of the most primitive plants, however, there is no discernible distinction between the sexes at all, even when magnified many thousands of times by means of a scanning electron microscope. Yet these apparently identical organisms copulate and produce offspring. Many examples are given by H. van den Ende in his remarkable book *Sexual Interactions in Plants*, which describes how in several different species the placing together of what appear to be identical and sexless individuals causes them to develop sexual organs which enable them to mate with each other. A period of getting acquainted and exploring each other is needed, during which sex hormones and other chemical messengers pass between them, leading to sexual arousal. It has been suggested that the hormone that causes sexual organs to develop should be called an *erogen* ("arouser of desire"), the hormone that attracts partners an *erotactin* ("creator of desire to touch") and the hormone that steers them towards each other an *erotropin* ("director of desire").

One striking case described by van den Ende is that of a green alga called *Chlamydomonas*. It is a tiny, single-celled organism which swims about in the water by means of a pair of whip-like threads known as *flagellata*. Every individual looks exactly like every other. What is more, there is no apparent difference between an ordinary vegetative cell and a sexual cell, whether male or female; in fact the two sexes are so impossible to tell apart by their appearance and structure that many scientists do not call them male and female but *plus* and *minus*. When an individual is ready for sex, it becomes transformed from its previous state, in which it showed no interest in the opposite gender, into a sexually mature being whose one aim is to find and copulate with another one of the opposite sex. The whole cell has been changed into a reproductive unit called a *gamete* ("marriage partner"), otherwise known as a sex cell. Its function is to join up with another gamete of the opposite sex so that their nuclei fuse together. The resulting cell, called the *zygote*, will become a new individual.

The copulation of gametes has often been described in *Chlamydomonas*, but it is only now beginning to be understood through recent advances in chemical analysis. The mating is a somewhat long drawn out affair, and is in two stages. First, in the species under investigation, when the two sexual types, positive and negative, are mixed together, they approach each other in pairs; if there are large numbers of individuals in close proximity they have a tendency to come together so that they form clumps of copulating partners sharing in a group experience. Beating their whip-like flagella in the water, the mating pairs approach closely. They touch the tips of their flagella together, seeming to be exchanging recognition signals. At this stage a certain amount of swapping has been observed. If the partners are compatible, the joining of the flagella proceeds, working from the tips towards their junction with the body of the cell. From between where the flagella arise, the contents of each cell begin to protrude, and as the rapidly lashing flagella continue to pair, both cells come into intimate contact and the contents begin to fuse together. After several hours of coupling the pair of gametes fuse completely to form a new zygote.

Experiments have shown that while the transformation of ordinary vegetative cells into sexual ones is largely, like the onset of puberty in any other species, a matter of age, feeding and life style, the attraction that draws the gametes towards their mating partners is a chemical one, brought about by secretions from their cells. In the species used in the experiments described by van den Ende the sexual cells are known as *isogametes*, which means "equal gametes," because in form and appearance they are identical and so cannot be distinguished as male and female; that, as we have seen, is why they are referred to as plus and minus instead. However, isogamy is an uncommon condition in the plant world and confined to some of the simplest algae and fungi. In most cases, however, there is a clear difference between male and female gametes. The male tends to be active and the female passive.

This difference is even known to occur in species where the male and female sex cells start off as isogametes, similar in form and equally mobile. A case in point is that of the brown alga *Ectocarpus siliculosus*, a very common seaweed around sea coasts, making large tufts with long, branched filaments. For some time after being released into the water the male and female gametes, alike in form and in activity, swim around without forming any sexual attachments; they seem not to be interested in each other in that way. Then something happens. The female gametes decide the time has come to settle down. They stop swimming around and fix themselves to a solid surface. Almost immediately, the male gametes, which up to then have shown complete indifference, become strongly attracted to the now motionless females. They cluster around the female cells in large numbers and attach themselves to them with their front flagella. It is not, however, that the male gametes prefer their females not to move; what causes the attraction is that as soon as they settle down the female cells start to secrete into the water a chemical substance which the male gametes find irresistible. However, it has been observed that the female cells do not settle down and start producing their alluring secretions unless there are male cells near; so it seems that the male gametes too must produce some substance that enables the females to recognize them. It has also been demonstrated that as soon as a female cell has been penetrated by a male gamete it stops producing its attractant, so any other male cells that have been attempting to copulate will immediately lose interest and go away.

Before we leave the seaweeds and the water, let us take a brief look at one of the fungi, an aquatic mold called *Achlya*. It may be considered out of place to include fungi in a book on the sex life of plants since, lacking chlorophyll and having a different type of structure, they are excluded by many biologists from the normal plant world. However, their sexual habits and techniques are remarkably varied and interesting, and sometimes call into question the assumptions we make about the fixity of male and female roles. *Achlya* is a fungus whose body is composed of *hyphae*, cylindrical filaments which spread both by growth at the tip and by branching, the whole network being called the *mycelium*. Within a species there are several strains, most of which are able to reproduce by self-conjugation, that is by producing both the eggs and the sperms which fertilize them. Some strains, however, behave in sexually opposite ways. When placed near each other so that they lie side by side in the water, male sexual organs called *antheridia* will start to appear on one and female organs called *oogonia* will develop on the other. Since it was not until the two individuals came into contact that the sex organs began to grow, it seems as if by physical contact with each other, by a sort of prolonged embrace, the partners have not only developed sexual desire for each other but actually grown the organs with which to consummate that desire. When the male organ has fully developed, it penetrates the wall of its partner's female organ and releases sperm cells which fertilize the eggs that have formed inside.

From studies that have been made of the entire mating sequence, it appears that several different chemical substances are involved. The whole sexual process is started by a hormone secreted by the female which on reaching the partner causes the male organs to start growing; the male in turn sends out a second hormone which starts sexual organs developing on the female; next a third hormone is produced by the female to attract and develop the male organs; then the male secretes a fourth hormone that stimulates the female organ to complete its growth and produce the eggs.

The most remarkable thing is that there appear to be varying degrees of maleness and femaleness in different strains of *Achlya*. At the extremes are thoroughly male and female strains that rarely conjugate with themselves but prefer a clearly opposite sexual partner. In between, however, are several grades which do not exhibit strongly male or female characteristics; they seem content to adapt themselves to the preferences of whatever partner they happen to find, and produce not only the right mating behavior but the right sexual organs to enable satisfactory sexual intercourse to take place. To quote van den Ende, "There is no commitment to a particular sexual character, since frequently a capacity to react either as male or female is exhibited depending on the sexual partner." Or to quote from J. R. Raper's *Sexual Versatility and Evolutionary Processes in Fungi*, there is "a sexual ambivalence in which maleness and femaleness are determined in each mating by common consent of the mated."

We must now come ashore and leave the algae and other aquatic organisms with which life, and sex, started. It was when the seaweeds or their predecessors began to move up the beaches, first only within the tidal zone and then beyond it, that the land habit was begun which after hundreds of millions of years was to lead to the higher plants that dominate the landscape today, and with which the rest of this book is mainly concerned.

The abandonment of the water brought considerable problems to plants in their sex lives. As we have seen, in the sea reproduction was a fairly simple matter; eggs and sperms were released into the water and could swim around till the males found the females, helped by chemical go-betweens to guide them. When the plants took to the land they had no water to transport the sex cells. How then was the male to reach the female?

Before we consider the answer, it is necessary to look at the nature of the sexual act and examine what happens when male and female gametes come together. All living things are composed of cells. The first and simplest ones were, as we have seen, single-celled organisms. After a time, some of them began to get together in groups, as in the so-called colonial algae, which are composed of individual and distinct cells partly joined together for cooperative living. The great advance came when multicellular organisms began to appear, consisting of many cells, often millions, no longer independent but part of the same organism, like a flowering plant or a human being. Different cells or groups of cells became specialized to perform particular functions, and so special structures came into being. Among these special structures are the sexual organs, and in most living things these are special in a way that makes them fundamentally different from all other organs, as we shall shortly see.

Every living cell of every organism except the most primitive contains a nucleus in which are the *genes*, the units of inheritance, carried on rod-like structures called *chromosomes*. The normal individual of any species, plant or animal, carries two sets of these chromosomes, one set derived from the male and one from the female parent. Because it has two sets it is called *diploid*. A normal human being has forty-six chromosomes, in two sets of twenty-three each, and a common daisy has eighteen, in two sets of nine. Sexual reproduction consists of a sex cell, or *gamete*, from each parent coming together to make a new offspring, the *zygote*. If, however, the number of chromosomes in each gamete were the same as in the body cells of the parents, that is to say two sets, then the offspring would have four sets. In the case of the daisy, each parent would contribute eighteen chromosomes and so the offspring would have thirty-six. After a few more generations of such reproduction, the offspring would have hundreds of thousands of chromosomes.

To overcome the problem and keep the number of chromosomes constant, there is a neat device. During the formation of gametes a process called *meiosis*, or "reduction division," occurs. By this means the number of chromosomes is halved from that of the body cells, so that there is only one set in each gamete, which is then called *haploid*, from the Greek for "single." Then when the male and female gametes come together at mating, each contributes one set of chromosomes and so the double set, the diploid number, is restored in the offspring. There is, however, one vital stage in the process of meiosis which must be noted. During the course of the reduction division the chromosomes perform a kind of intricate ritual dance in which occurs what is known as "crossing over." This means that each chromosome joins up with its counterpart in the opposite set, and they then exchange parts of their body with each other, so that when they finally split apart into single sets the genetic material has been shuffled, part coming from the male grandparent and part from the female one.

In this way, by a recombination of genes that is different for every gamete, the material of heredity is kept continually shuffled and redealt, so perpetuating the differences between offspring which are the whole purpose of sex.

Having taken a look at the way gametes are formed, we can now return to the question of how the male reproductive cell solved the problem of reaching the female when plants took to the land and could no longer simply mate in the water. It took many millions of years to answer the problem, during which time the plants that had come ashore were not able to venture far onto dry land because the male reproductive cell still had to swim to reach the female and so needed at least a film of water along which to travel.

Several of our present-day land plants remain at the stage where they need damp conditions for fertilization because the sperm has to swim. Among them are the mosses and the ferns, both of which show two quite different phases in their lives, one sexual and one nonsexual. If you examine a fern you will find that around the edges of the leaflets, or underneath them, according to species, are structures called *sporangia*, which, when they are ripe, split open and release vast numbers of tiny spores.

Spores are tiny units of reproduction and distribution evolved by early forms of life long before the higher plants arrived on the scene. Indeed spores came into being before sex was invented, and remain the means of distribution of all manner of organisms. After the fern spores have fallen on the ground they will, if the conditions are right, germinate and grow. But what grows from the spore is not at all like a fern plant in appearance or structure. It is small, flat and heart-shaped, and easily overlooked on the surface of the soil. Called the *prothallus*, it represents a distinct stage in the life cycle of a fern. After a time sex organs develop on the prothallus; from the male organs sperms are emitted which are able in the moist conditions close to the ground to swim to the female organs and fertilize the egg cells in them. The male and female gametes have undergone meiosis during their formation, so that each has only one set of chromosomes and when they have mated they produce new offspring with a double set of chromosomes. This then grows into the diploid plant that is recognized as a fern. In turn that produces spores which fall to the ground and germinate into the tiny, heart-shaped prothalli, and these develop sexual organs, which produce eggs and sperms, which mate and produce new fern plants, and so the cycle goes on.

This process is known as the *alternation of generations*, and it is the essential life cycle of all plants beyond the most rudimentary and primitive. The two phases are quite clear and distinct: the *sporophyte*, a nonsexual stage producing spores in due course, and the *gametophyte*, which carries the sexual organs in which are produced the haploid gametes.

Usually it is the sporophyte generation that we recognize, as in the case of the ferns and of all the flowering plants. In some cases, such as that of the mosses, however, it is the other way round. The cushion-forming plant which we know as a moss is the gametophyte, bearing the sex organs. The generation that results from sexual intercourse is the capsule which is borne on a stalk above the plant and in which the spores are formed, to be released in clouds on the wind when mature and produce new moss plants. The capsule is the sporophyte generation and lives in a parastic way on the sexual generation, the gametophyte.

Because their sexual and nonsexual stages are so distinct in appearance, the alternation of generations is very clear in the case of the ferns and the mosses, as it is with many other of their relatives among the cryptogams. Since they do not bear seed, they are reliant on spores for their dispersal. Though this can be highly efficient, spores being so minute and light that they may be carried hundreds of miles on air currents, it does limit their evolutionary scope and their capacity for sexual expression in a variety of forms. This, as we have seen, is basically because the male reproductive cell needs a watery medium to be able to reach the female.

The seed plants, and particularly the flowering plants, have solved the problem so brilliantly that they are able to indulge in sexual intercourse in almost any conditions. They still have a life cycle with an alternation of generations between a sexual and a nonsexual phase. But in their case the two generations are not easy to tell apart. The plant itself bears the spores, and these spores are of two kinds: *microspores* and *megaspores*. The microspores are the pollen grains made by the male sex organs and producing the male gametes. The megaspores are hidden inside the ovaries of the flowers, where the ovules are found, and in those ovules are the megaspores containing the female gametes.

During the process of evolution the male microspore has remained small and mobile, while the female megaspore has become larger and more passive. It is with the getting of that male to the female that sex, and this book, are concerned.

IV

The Sexual Organs

Flowers are sexual organs dressed up to appeal to insects. That is true of the majority of higher plants, of which over a quarter of a million differing living species are known, with new ones being discovered all the time (and some become extinct, mainly because of human activity). By comparison, there are estimated to be between half a million and a million different kinds of insects, all of them in the end, directly or indirectly, dependent on plants, though not all of them are immediately involved in their sex lives. There are exceptions, too, on the part of the plants; in some species the flower attracts different creatures, such as birds, bats or other animals, and in some it does not need to make itself attractive at all, because it relies on wind or water for pollination.

These last two methods, however, are thought to have come later. It may on the face of it seem strange that pollination in the water should be a recent development among the flowering plants, since, as we have seen, sexual intercourse first took place in the ocean, long before there were any plants on dry land to indulge in it. But those primitive forms of plant life that invented sex in the water were very different from the higher plants with which this chapter is concerned. They had no flowers and no pollen, so what went on between them sexually as they bathed together could not be called pollination, however varied and satisfying it may have been. The true flowering plants that use the water for the purpose of sexual union show every sign of having been descended from land plants and having adventured into the water in recent times. We will look at some examples of this *hydrophily* (literally "love-in-the-water") later. It is rather rare in the flowering plants, probably because pollen was designed for dry conditions and finds the water uncomfortable and dangerous.

Anemophily ("wind-love") is much commoner and more important. Wind is the normal pollinating agent of the grasses, the sedges and the rushes, also of a great many trees. The golden clouds of pollen carried on the breeze every spring contain millions of grains (and start the annual misery for hay fever victims, for whose benefit pollen counts are made throughout the season in many countries). The rate of pollen production is prodigious; dedicated investigators who catch the grains on special slides and then identify and count them have estimated that even in a fairly urban site there is a "pollen rain" each year of a thousand grains from trees and two thousand from grass per square centimeter. Those figures were for Great Britain; in more thickly forested parts of Sweden it is estimated that the yearly fall of pollen may amount to as many as thirty thousand grains on each square centimeter.

Because of the extravagant production of pollen involved in pollination by wind, it was thought by many botanists until fairly recently to represent a primitive state of affairs. Since such wholesale broadcasting, in the hope that some of the grains would fall by lucky chance on a receptive female surface, was exceedingly wasteful, surely it must have come before pollination by insects, a much more precise way of getting pollen from the male to the female. Because it relied on the wind, a tree such as the hazel had to produce from each catkin about four million pollen grains, of which no more than a tiny few had any chance of reaching a female and having young. How much more economical and efficient it would be to have the pollen carried by an insect, going methodically from flower to flower instead of being blown everywhere at once. And how much less pollen would be needed.

Besides, it seemed quite obvious that as wind had been around long before there were any insects, and had been the only thing other than water able to act as a sexual go-between for the primitive forms of plant life in those times, it must have come before insects as the pollen-carrier in the flowering plants too.

The argument seemed so reasonable that it was accepted by many botanists for a long time, and it still crops up in books on nature study to this day. But like so many reasonable arguments it suffered from one serious defect. It was based on theory instead of observation. There was a kind of moral assumption about it that waste was wrong, particularly waste of male sperm, the loss of which is so often regarded with superstitious fear as weakening.

Current botanical opinion is that the assumption is false, and that in the case of the flowering plants, the *Angiosperms*, wind came well after insects as an instrument of pollination. There are many reasons for thinking so. The wind cannot smell or taste; yet several wind-pollinated plants still produce scents and flavors, with which they or their ancestors most likely used to lure insects before they finally threw their sex lives to the wind. The common stinging-nettle, in spite of being wind-pollinated, has in its flowers glands designed to produce nectar, which insects delight in, but which holds no interest for the wind. Similar nectaries are found in the nettle's near relation *Cannabis*, which may have drugged insects into a pleasurable stupor long before human beings started using it to drug themselves.

Then there are many cases where among very closely related species, even members of the same genus, some rely on wind and others on insects to pollinate them. That annoying lawn weed the common plantain, *Plantago major*, has unattractive, dirty white flowers, as unlikely to appeal to insects as to human beings; careful observers report that the flowers are pollinated only by the wind. Yet the very similar Hoary Plantain, *Plantago media*, has beautifully colored little flowers with an attractive scent, and is regularly visited by several insects, including hive bees, which are quite sophisticated in their tastes and demand a high standard of attractiveness in their floral partners. A third species, the Ribwort Plantain, *Plantago lanceolata*, was thought until recently to be entirely pollinated by the wind; then biologists began to wonder about the many visits that hoverflies were seen to make to the flowers. Hoverflies are well-known pollen-eaters. Did they only rob the flowers, or did they perform any useful function in return for their meal? After some careful studies of the flies and their diet, followed by detective work making use of artificially dyed pollen grains, it was found that the hoverflies did indeed do a useful job of carrying pollen from one plant to another.

An opposite example is given by the common heather, *Calluna vulgaris*, which is clearly designed to attract insects, with its rose-purple flowers and the plentiful nectar from which bees make delicious honey. Yet heather produces great quantities of fine pollen, which is carried by the wind and must fertilize a large proportion of the flowers. Such a surplus of pollen production goes way beyond the amount which would be necessary for pollination purely by insects, and demonstrates again that where their sexual needs are concerned many flowers like to hedge their bets and have more than one type of mating agent. And there are cases of flowers that are not wind-pollinated at all yet produce pollen on an even larger scale, such as the ordinary field poppy *(Papaver rhoeas)*, which has been shown in many studies to produce far more pollen than most wind-pollinated plants.

In any case, the argument that waste must be evidence of a "lower" stage of evolution has been shown to be completely untrue. Many of the most highly evolved forms of life of the present day are prodigiously "wasteful" in their methods of propagation, squandering millions upon millions of male reproductive cells for each offspring produced. Human beings themselves – considered by most biologists to represent the highest peak of evolution so far achieved – are so wasteful in their method of reproduction that each time a baby is conceived about three hundred and fifty million spermatozoa are thrown away and only one reaches its target.

Since wind pollination seems to have come rather late in the history of the flowering plants, we will leave it till later in the chapter and deal first with pollination by insects. It was on that relationship that the whole development of flowers depended; when we admire a beautiful rose we are only showing that we share some of the tastes of the insects, for whose sake the rose's attractions were intended. However different we may feel ourselves to be from the insects, we seem to be drawn to flowers for some of the same reasons as they are: form, color and scent. We even, as we shall see in more detail later, use flowers as many insects do, for our own sexual purposes.

Entomophily (from the Greek for "insect love") is the basis on which the flowering plants have been founded. A hundred million years ago, when flowers are thought to have started making their appearance, there were already vast numbers of different types of insect; their fossil remains show that some of them were very like kinds still living today. Most of them were beetles, though there were also plenty of bugs, thrips and flies. Like the more primitive of their relatives alive today, they were no doubt crude creatures of limited intelligence, and would have been quite unable to manipulate the complicated mechanisms which many modern flowers have developed for their sexual purposes. But those dull-witted creatures were ideally suited to the earliest flowers, which were extremely simple in their structure and require no psychological subtlety from their visitors.

The fossil record shows us that many of these early flowering plants were remarkably like some species that are still alive today. And they are still visited by primitive types of beetle and suchlike backward insects. Since these earliest kinds of flowers are also, as might be expected, the simplest, perhaps this is the point at which we should examine the parts of a typical simple flower to see what they are for and to learn the floral facts of life. Basically there are four sets of parts to a "complete" flower. Two of them are essential: they are the sexual organs, and without them – or at least one of them – no normal flower would have any use or purpose (though there are sometimes sexless flowers whose job, as we shall see, is to advertise the sexual availability of others). The remaining two are not directly involved as sexual participants, so they are dispensable, and indeed often dispensed with; because of their minor role they are sometimes called the "accessory parts."

The female organs, taken together, are called the *gynoecium* (from the Greek meaning "women's quarters") and are usually in the center of the flower. They consist of one or more *carpels*, each made up of a hollow "womb" at the base, called the *ovary*, topped with a stalk-like projection, the *style*, ending in the receptive female surface called the *stigma*. Inside the ovary are the *ovules*, containing the egg cells, which when fertilized by male reproductive cells will develop into seeds.

The male organs are together called the *androecium* ("men's quarters") and are made up of *stamens*, each consisting of a bag of pollen, called the *anther*, usually carried on the end of a stalk called the *filament*.

When the ovules have reached the stage of development at which they are ready to be fertilized, the stigma becomes sexually "ripe." Secretions from the female tissues prepare the way for sexual intercourse, in a similar manner to the secretions produced by a female animal when ready for mating. The surface of the stigma may become moist, and often sticky, so as to trap any pollen grains which might find their way onto it. The sticky secretion is usually found to be sweet to the taste, and analysis shows that it contains a considerable amount of sugar, varying in concentration according to the species of plant.

If the pollen is compatible with the stigma, sexual union can begin. Compatibility is, however, as in all sexual matters, quite a complicated thing. Each partner must find the other agreeable before anything can happen. In the first place they must be suited to each other in their actual physique, so that they can fit intimately together. It is of little use for big, coarse-grained pollen to attempt full physical intimacy with a delicately surfaced stigma; indeed, as we shall see in the chapter on sexual taboos, physical incompatibility of this kind is used by some species of plants to prevent unsuitable mating. It may be thought easier for a smaller and more finely textured pollen grain to fit comfortably into the indentations of a coarsely surfaced stigma, but in most cases there does seem to be a need for close physical contact between as large an area as possible of the two surfaces, so that their microscopic bumps and hollows fit. If the pollen only touches the stigma casually at one or two points, it appears that further development of the relationship is less likely; more intimate contact seems necessary for the union to be taken seriously.

Besides physical compatibility, the chemistry needs to be right, on both sides. There is a period of foreplay, during which the female organs send and receive chemical messages to tell them whether the pollen is an acceptable sexual partner or not. At the same time, the pollen will be "tasting" to decide whether the female secretions on the stigma are likely to lead to a satisfactory outcome. Pollen from differing species of plants has very distinct preferences as to what it needs to stimulate it into action. In particular different kinds react differently to degrees of sweetness. Most types of pollen can be germinated artificially in a solution of sugar in water, but they differ widely in the way they perform in different concentrations. Pollen of a species of senna *(Cassia)* will germinate happily in a seventy per cent sugar syrup, but most kinds could not put up with anything like that strength. The pollen of the snowdrop (*Galanthus* species) germinates best in a solution with only one or two per cent sugar, and many kinds germinate in pure water. Some pollen will stubbornly refuse to germinate at all except on the stigma of its own species.

Such individual requirements are no doubt very useful in preventing many kinds of pollen from wasting themselves in germinating before they reach the right stigma. The result, as in many human encounters, is that the degree of sweetness that one male finds attractive may turn another off as being much too sugary.

Let us suppose that there are no problems of incompatibility and that the right pollen grain has been deposited by some means on the right stigma: ripe, receptive and sweet, but not too sweet. Pollination could now be said to be complete, if by the term pollination is meant simply the bringing together of a compatible male and female pair. That is, as we shall see, the main task that flowers exist to perform, and to which they devote amazing ingenuity. But introduction, however well arranged, is only the beginning, and is intended to lead on to other things.

Pollination, therefore, is a word used by most biologists to mean not just the introduction but also the complete sexual act which follows. The union is not consummated until the male has penetrated the female far enough to release a reproductive cell in such a position that it can reach an ovule to fertilize it. Fertilization may or may not then take place; sexual intercourse does not necessarily lead, in the vegetable as in the animal kingdom, to pregnancy.

The way in which the pollen penetrates the female organs varies slightly from species to species, but the basic principles remain the same. As we have seen, given the right secretions and encouragement from the stigmatic surface the pollen will germinate, rather like a seed. A protuberance will start to grow out from the grain and force itself down through the stigma and into the style, driven by two forces: to get away from the air and to find its way towards the source, hidden deep in the female organs, of the sugars and other secretions being sent through the tissues. This protuberance is called the *pollen tube*, and it pushes its way down through the style towards the ovary. If the style is a long one, the tube may have to travel a considerable distance for something of such microscopic diameter. In pushing through the style, the tube may be helped and guided in some species by special canals, or the cells may be loosely arranged so that the tube can thrust its way between them without meeting much resistance; in other cases the tube may receive little or no help and have to penetrate through the cells themselves, destroying them and creating its own channel as it goes along.

When the pollen tube reaches the hollow of the ovary, it makes its way to an ovule, which it enters through a tiny, cervix-like channel called the *micropyle*; this is what happens in nearly all cases, but a few pollen tubes prefer to get in through the rear of the ovule. When it has squeezed its way into the ovule, the pollen tube releases two male reproductive cells, called *gametes*. These, as we have seen, have gone through reduction division, or *meiosis*, and have only half the number of chromosomes as there are in the plant's ordinary cells.

Fusion of one of these male gametes with the *egg cell*, or female gamete, in the ovule may follow. If so, *fertilization* will have been brought about, but that is a different process and will be dealt with in a later chapter. The act of *pollination* has been completed when the pollen tube reaches the ovule.

Having dealt briefly with the nature and function of the two essential parts of the flower, the male and female sexual organs, we can now examine the other two structures, the "accessory parts," present in a very large proportion of flowers. These parts may play a minor role in the immediate sex life of most flowers, but they play a major role in helping to create the right type of blossom to attract the insects and other creatures on which pollination depends.

The first of these accessory parts is the external one called the *calyx*. In most species it is the earliest visible sign that a flower is beginning to form. It makes up the green exterior of the flower bud, and its primary function is to protect the young organs inside during the time when they are developing and before they reach sexual maturity. The calyx may be made up of separate segments called *sepals*, as in the buttercup *(Ranunculus)*, in which case it is *polysepalous*, or the sepals may be united, as in the primroses (Primula), in which case they are called *gamosepalous*. Sometimes the function of the calyx is so exclusively to protect the developing organs that it becomes useless and drops off as soon as the flower opens, as in the poppies *(Papaver)*. In other cases it may stay on and perform some quite different function, as we shall see.

The second set of accessory parts, normally placed between the calyx and the sexual organs, is the *corolla*, and its role in life is to advertise. It may be made up of separate *petals*, in which case it is usually called *polypetalous* ("many-petalled"), though a more precise term is *choripetalous* ("separate-petalled"). On the other hand, the petals may be joined together from the base upwards, in which case the flowers are often called *gamopetalous* ("with married petals"), though since the joining of the petals is not sexual the term *sympetalous* ("with united petals") is preferred by purists because it has no sexual overtones.

Taken together, the calyx and corolla may be called the *perianth*, and its parts the *perianth segments*. When they all look alike, so that no distinction can be drawn between them, as in the tulip, the segments are called *tepals*, from an anagram of the word petals.

The flower is usually carried on a stalk called the *pedicel*, and the end of the stalk, where the flower is attached, is called the *receptacle*. Sometimes this hardly looks any different from the rest of the stalk, but more often it is thickened. Something it is greatly swollen, as in the strawberry, where the juicy part is the enlarged receptacle, made up of distended stalk tissue, and the real fruits are the pips on the outside. There are very many different ways in which the flower can be attached to the receptacle; some are the opposite extreme to the strawberry and have all the lower parts of the female organs embedded deeply in the receptacle, so that the ovary is entirely surrounded by it.

When the base of the female organs is completely enclosed by the receptacle in this way and the other parts of the flower grow from the top of the receptacle, the ovary is said to be *inferior* and the flower to be *epigynous* (Greek for "on top of the woman"). When the female organs sit on the top of the receptacle and the rest of the floral parts grow from below them, the ovary is said to be *superior* and the flower to be *hypogynous* ("underneath the woman"). There is a third type where the female organs are set at the center of a shallow basin formed by the receptacle, and the other floral parts are borne around the rim of the basin; in this case the ovary is still said to be superior, and the flower is *perigynous* ("around the woman").

Those are the three basic positions. There are all kinds of variations possible, and in the case of many different species of flowers with their parts in intermediate positions it is not really possible to say whether the female organs are on top or underneath. The classification by female position is bound to be somewhat arbitrary in these cases, but it is very useful in helping botanists to identify species of plants. And it has been of the greatest possible importance in the study of the evolution and development of the flowering plants ever since the earliest traces that can be found in the fossil record.

It is generally agreed that the most primitive floral position was hypogynous ("woman on top"). This is the pattern found in the magnolia family, which is thought to contain some of the earliest types of flowering plants still alive today. The female organs, consisting of many separate carpels arranged on a long receptacle, protrude well above the large and indefinite number of stamens in a magnolia flower, and the petals and sepals are similar to each other and separate *(choritepalous)*. Most botanists believe that this type of flower developed from earlier types that consisted of a string of female organs along the end of a shoot, with male organs coming behind them, and behind that ordinary leaves along the stem. Beetles wandering along the shoot in search of food would find the pollen greatly to their liking. They and their relatives had, for many millions of years, been enjoying the somewhat different kind of pollen produced by the long-established *gymnosperms*, that more ancient section of seed plants which still flourish in the form of pine trees and other conifers. Being crude creatures with messy eating habits, the beetles would crawl over both the male and the female organs, smearing pollen from one to the other and so bringing about pollination.

Over the course of millions of years those early structures would give rise to new and improved forms, more like the flowers we know today. The receptacle would become shorter, and the female organs along it would be more bunched together, to increase their chance of all being crawled over by the pollen-smeared beetles. The nearest leaves, instead of remaining strung out along the branch, would grow closer together, so as to keep visiting beetles in one place till their pollinating job had been done instead of letting them wander aimlessly about the shoot. The bunch of leaves would become different in color and texture from the rest and so attract the attention of the rather short-sighted beetles, and then would evolve into petals, or perianth segments, as in present-day flowers. In that way the magnolias and their primitive relations would have developed millions of years ago into the form which has remained more or less unchanged ever since among their living descendants. And the same kinds of primitive beetles still visit them and pollinate them in the same crude way, what botanists call the "mess and soil" method.

Since their early days a hundred million or so years ago, the flowers have developed an astonishing diversity of forms, colors, shapes and sizes, in close partnership with the insects, which during the same period have evolved into an equally amazing variety of different creatures. Even from the ranks of the stupid primitive beetles and flies have come some much more talented descendants. And totally new forms have arisen of much greater sophistication and intelligence, such as the modern bees, which possess such powers of discernment that they could be said to have finer feelings.

In developing their relationship with the increasingly complex insect world to fulfill their own sexual needs, the flowers have had to play what used to be called a traditionally feminine role. Since they remain fixed to one spot, they cannot take the initiative, or at any rate appear to do so. They must sit and wait. It is the insect that, as the mobile partner, makes all the advances, or at least behaves as if it does. The flower has to be subtler in its approach, making itself as attractive as possible but having to wait upon the pleasure of visitors, who can take it or leave it alone.

The simple magnolia flowers are just right to appeal to the tastes of backward and clumsy beetles, which trample over them chewing pollen and dropping bits about the place, but more sophisticated visitors like bees find such crude flowers unattractive; to quote Professors Faegri and van der Pijl, bees are ill at ease in these large, loose blossoms "like a small child in a big bed." The history of floral development has been one of constant experiment with new forms to appeal to ever more intelligent pollinators. From those primitive flowers with separate petals and a superior gynoecium ("woman on top") have developed over the years the most modern forms, designed for the preferences and abilities of the most modern visitors, with the perianth segments joined together, often brightly colored, and the gynoecium inferior ("woman underneath"). Some of the most highly evolved species have developed such specialized structures and devices that only one pollinator has the capacity to satisfy them. This makes for much greater precision, since the pollinator can be relied upon to hit the right spot every time, with no waste of time or effort. On the other hand, it puts the species at great risk, because if the only creature able to satisfy its needs should for some reason not be available, or be driven from the region by human activity or some natural disaster, or even become extinct, that plant will have no other mating agent to turn to and so may die out. We shall see later, in a slightly different context, what has happened to a species that used to rely on the dodo before that bird was hunted to extinction. At least the more primitive types of flowers are not likely to suffer that fate, because they are open and available to any and every kind of creature that cares to visit them.

Apart from the physical devices that flowers have developed to lure and ensnare pollinators, including drugs and traps, there are the advertising methods they have perfected to catch the attention of potential visitors. One of the oldest attention-getters is, as we have seen, scent, which can in suitable circumstances arouse the interest of a prospective partner at a considerable distance; it is the equivalent of an advertisement in the personal columns saying "Attractive, willing flower, newly opened, seeks suitable pollinator for sincere relationship." For the purposes of this chapter, however, we will not treat scent as advertising, because it often provides the visitor with real satisfaction, since scent forms an important part of the taste of pollen, nectar or other food substances which the visitor takes from the flower to feed itself or its young. In some cases, indeed, scent appears to be a complete satisfaction in itself, the insect concerned apparently being quite happy with smell alone and not wanting anything else from its visit.

We will use the word advertising only for things designed solely to attract attention and create desire, not for things that actually satisfy an urge. The most obvious piece of pure advertising in that sense is, as already noted, the display made by the petals, or sometimes by the sepals, or even by all the perianth segments together, using color, texture, form and brightness to catch the eye and make a favorable impression. Such display is designed for exactly the same purpose as bright packaging around goods in the shops or fashionable clothes around a person: to capture attention and to arouse interest in what is inside.

A great deal of study has been devoted to the chemical substances that create the colors in plants. The subject is a complicated one and by no means everything is known about it, but there seem to be two types of substances involved: one type is generally responsible for the purples, blues and most reds and the other largely for colors in the yellow part of the spectrum. Together or separately, and in conjunction with other substances in the cells, they give the whole range of colors that can be seen in the world of flowers. Pure white is generally caused by light reflections in the spaces between cells without pigment, and black – though there are no really pure black flowers – to reflections between cells with complementary colors.

Since colors are so important in an advertising sense, they need to be as clear and attractive as possible when the flower is ready to receive visitors. Some colors, however, are not fade-proof and will go pale in the light almost as soon as the flowers open; they need to catch a visitor's eye quickly, before their charms have gone. Many flowers change color after they have opened; the Lungworts *(Pulmonaria)* start pink and then turn blue, and their relation the Changing Forget-me-not *(Myosotis discolor)* has flowers that open yellow, then turn red and then turn blue. The reasons for this odd behavior are not fully understood, but it is suggested that it is so that the flowers can attract visitors at just the right time, not before or after they are ready.

One very neat use of change of color is shown by the Horse-chestnut *(Aesculus hippocastanum)*, which has a rather complicated sex life. It is *andromonoecious*, that is to say it has separate male and hermaphrodite flowers; the male flowers are sexually ready first, but the hermaphrodite ones start female. It is an odd sexual arrangement, but it works remarkably well, to judge by the large number of seedlings produced every year. The use of color in this case is to tell the bumblebees which flowers to visit and which to leave alone. What happens is that when the flowers open the white petals are marked with yellow spots at the base to guide the bees to where the nectar is, so that they may find it easily, become powdered over with the reddish pollen as they drink, and then pass on to the next blossom with as little delay as possible. When a flower has no nectar left, the yellow spots turn red. Just as during their yellow period the spots were very attractive to bees, one of whose favorite colors is yellow, so when they change to red they completely lose their appeal. It is not so much that bees dislike red as that they cannot see it. They are red-blind, so the changed spots do not look red to them but black. Observations have shown that most bees ignore the flowers marked with what to them is black and go for the younger ones still wearing yellow. The change of color is a labelling device to stop bees from wasting their time on flowers that are no longer young. It may only knock a second or two off the time spent at each blossom, but with a whole tree, carrying thousands of flowers, the saving is considerable; it could make all the difference, in an unfavorable season when the blossoming period is short, between a good and a poor crop of fruits.

A great deal of study has been made of the colors that are most likely to appeal to different types of visitors. On the whole it seems that most of them are very conservative indeed, preferring the same old colors which appealed to their parents and grandparents. Occasionally a new shade will appear in a blossom or two, particularly among garden flowers, which because of the artificial life they lead tend to be unstable; but the more strikingly different the more likely it is to be avoided by the usual run of insects in favor of the more conventional colors of the rest of the flowers. Very rarely the new shade will strike a chord in some particularly responsive visitor and so a new fashion may be started, but the odds are heavily against its success. Of course, if the visitor attracted by the new color is a human being in the commercial plant business, he will attempt to propagate from the new plant as quickly as he can, in the hope of making a fortune by launching it in thousands or millions as a "breakthrough." But insects do not much care for breakthroughs; they have spent millions of years of evolution building up a relationship with one color or set of colors and so are unlikely to fall for the latest novelty.

Thousands of experiments have been carried out by naturalists, amateur and professional, to try to discover how insects react to different forms and colors. Because of their importance in pollination, their intelligence, their ability to learn and to remember, and the ease with which they can be kept and observed, hive bees have been used in more of such experiments than any other insects. A great variety of techniques has been used, from simple observations in the wild to highly elaborate laboratory methods in totally controlled conditions. Real and artificial flowers have been used; also flat cardboard shapes, from simple circles and squares and stars to complicated geometrical designs, in white, black, grey and every imaginable color and combination of colors, including ones beyond the range of human vision. Three-dimensional models have been constructed of varying degrees of complexity. Sugar-water, honey and other sweet substances have been added or withheld. Natural and artificial scents have been used, some pleasant and some disgusting (to human beings, that is, though some insects would rate them the other way round). Different patterns of cage have been tried, lit by natural and artificial light of every color in, and beyond, the visual spectrum, in which bees could be offered any of these inducements, alone or with others, to see which they preferred. Students of insect behavior have used all these, and other, devices to measure response to different stimuli and published their findings in learned papers. Some remarkable discoveries have been recorded which, if not always entirely consistent with each other, did put our knowledge of insect responses on a firm footing.

Among all this research, one series of experiments has remained a classic ever since the results were published in 1956. Using a specially built "spectral color-mixing apparatus" of advanced design, Daumer was able to compare the reactions of bees to light of every color at varying degrees of intensity. Many of his findings were not new, but the painstaking care with which he had conducted and recorded his investigations enabled Daumer to present the facts with new thoroughness and precision.

The technique of training bees to alight on a paper of one particular color to feed, ignoring other papers of different shades of grey, even though food was equally available, had already been developed many years before by students of insect behavior such as von Frisch, and elaborated by many other researchers. From such studies it had been established that bees had surprising powers of being able to pick out a special color and show great loyalty to it. Using refinements of basically similar methods, Daumer trained bees to respond to a particular color and then set about finding how far they were able to distinguish that color from others, both near to it and far from it in the visible spectrum. As a result, we can now look at colors from the bee's point of view, which is not at all like that of a human being.

As we noted in the case of the horse-chestnut blossoms, bees cannot see red at all. To them it does not exist as a color. So any flower that was pure red would be entirely wasted on them. To make up for that, however, at the other end of the visual scale bees can see something that is totally invisible to us, namely ultraviolet light. So both bees and humans are denied, though at different extremes, part of the ordinary daily visual experience of the other. And even those colors we can both see must appear very different to them and to us.

Red, with the longest wavelength of all colors visible to the human eye, is seen by most people as the "hottest" color, full of excitement and danger and sexual stimulus. That is no doubt why a bunch of red roses means "I love you," why there are red light districts, and why people talk about scarlet women.

To red-blind bees, though, yellow is the color with the longest wavelength, so it must be to them what red is to us. Blue, which humans see as cool, must appear a middling sort of color to a bee, much as green is to us. The really cool color to a bee is presumably ultraviolet, which we cannot even see. That is why in appealing to bees – and to all the other insects which have the same color-vision – flowers have to use colors that are quite lost on us.

Unfortunately we shall never, even with all the knowledge we now have on the subject and with our most advanced scientific instruments, be able to see the world the way the bees see it, simply because our eyes are not equipped to do so. There do seem to be reasons, though, for suggesting that we lose more by being blind to ultraviolet than bees do by being blind to red. There is little doubt that they can not only distinguish between colors better than we can but actually see a number of different colors. That is partly because their many-faceted eyes are quicker than ours to respond to the changes that occur all the time because of differences in light intensity caused by the time of day and the interplay of sunshine and cloud. Much more important, however, is the fact that the colors are more spread out and separated in the bee's visual spectrum.

To understand the differences between a bee's color vision and our own, let us suppose that a color television had to be designed for bees instead of humans. It would have to be planned on new lines. The screen of an ordinary present-day television set meant for human viewers is covered with tiny dots, each of which shines red, green or blue. There are equal numbers of these three different dots and they are spread evenly over the surface. From the varying amount of light shining through those dots a complete picture is built up, faithfully reproducing any color that the human eye can see. The reason for the choice of red, green and blue is that they correspond to the three different kinds of cone in the normal human eye: red-sensitive, green-sensitive and blue-sensitive. On these three receptors rests the whole of our color vision; every other shade that we can see is made up from a combination of stimuli to the three kinds of cone. That is why red, green and blue are called the *primary* colors of the spectrum.

A picture on a television set designed for humans would be a complete disappointment to a bee, because a third of all those dots, the ones that shine red to us, would convey no color at all to it. The dots on the screen of a television set designed for bees would have to be yellow, blue and ultraviolet, because those are the colors to which the three kinds of cone in the eye of a bee are sensitive. In fact there are more ultraviolet-sensitive cones than either of the other two kinds, so perhaps a television for bees should have more ultraviolet dots than yellow or blue ones. Such an arrangement of dots would give a bee a perfectly satisfactory picture, containing all the colors the insect could ever possibly see; but the picture would be just as disappointing to us as that on a human's television set would be to a bee, because all the ultraviolet dots would appear colorless.

Among the remarkable facts established by Daumer's researches was the ability of bees to distinguish accurately between their primary colors – yellow, blue and ultraviolet – not only when brightly lit but in conditions of such dim illumination as to be almost total darkness. A large number of different shades of each bee-primary color was tested, and it was found that bees could tell some yellows and some blues apart; but in these cases the ability to distinguish between similar colors fell off as the lighting was reduced. In the ultraviolet waveband, however, bees were able to recognize differences between certain wavelengths even in the most subdued light. In addition to their primary colors, bees were perfectly capable of recognizing secondary colors; between yellow and blue they could distinguish the color blue-green (what photographers call cyan) and between blue and ultraviolet they could see the color violet, which is at the extreme of the human range of vision.

In addition to ultraviolet itself, there are two other colors that bees can see but humans cannot. Since they are the equivalent from a bee's point of view to what we call white and purple from our point of view, they are usually referred to as "bee-white" and "bee-purple." A white flower is one that reflects all the primary colors. To a human that means red, green and blue; but to a bee such a combination would not look white at all but colored, because it lacks one of the bee's three primary colors, ultraviolet. So "bee-white" must contain ultraviolet. In the same way, purple is made up of a mixture of the primary colors with the longest and shortest wavelengths. To a human that means red and blue; but to a bee a mixture of those two colors would merely look blue, because the red would be invisible. So "bee-purple" must be a combination of yellow and ultraviolet, which merely looks yellow to us.

In being unable to see bee-white and bee-purple, it seems that humans are not missing just two colors but many more. This arises from the fact, already touched on, that the eye of a bee is extremely sensitive to different amounts and wavelengths of ultraviolet light. In one series of experiments it was shown that a mixture of as little as two percent ultraviolet to ninety-eight percent yellow enabled bees to distinguish it completely from pure yellow; on the other hand no less than fifty percent yellow had to be mixed with ultraviolet before bees could tell it apart from pure ultraviolet. It is clear from these and other findings that both the proportion and the wavelength of the ultraviolet component play a dominant part in determining which of several different possible hues of bee-purple is seen by a bee; and there must similarly be different shades of bee-white, varying in their impact on a bee according to the nature and amount of the ultraviolet content.

We have spent some time on the color vision of bees, because more experiments have been carried out with them than with other insects. That is partly because of the commercial importance of bees to humans, not only as manufacturers of honey, which has become big business in many parts of the world, but as pollinators of economically important crops. We have become increasingly dependent on the domesticated hive bee or honeybee, *Apis mellifera*, which is kept on a large scale by people in every part of the world with a suitable climate. Ecologists are alarmed at the artificially maintained population explosion of honeybees, because it is upsetting the balance of nature and endangering the continued existence of many species of wild plant and wild bee which rely on each other for survival. Such relationships, built up through the ages, can be only too easily destroyed.

The chief reason for the use of the honeybee in experiments, however, is that it is the ideal laboratory subject. It is a member of a highly organized community run on totalitarian lines; each bee knows exactly its place in the social order and would never question that place or attempt to move out of its class. Its behavior is always predictable, and it has no personality problems because it has no personality. It can be taught, it can remember its lessons, and it is obedient; in short, it is what laboratory workers call "intelligent." And because of its capacity for hard work, its total conformity and its utter subordination of itself to the good of the community, the honeybee has for centuries been held up as a shining example to mankind, as in those well-known lines by the hymn writer Isaac Watts, in *Divine Songs for Children*, "How doth the little busy bee improve each shining hour, and gather honey all the day from every opening flower!" So there was a tendency, as we have noticed in other contexts, for certain moralistic assumptions to be made that in some way studies of bees' behavior would provide lessons for human conduct. Earlier observers, as Faegri and van der Pijl point out, presumed that the senses of pollinators corresponded to those of man. Today we know that they do not necessarily do so; and although in modern experiments it is often astonishing to see how similar are the effects of sensory activities in insects and in man, there are also obvious exceptions, like different ranges of sight, touch and smell. One serious source of experimental error, Faegri and van der Pijl add, is that both the experimenter and his animals ultimately become too clever, and the latter may be trained to perform tricks never performed under natural conditions.

One powerful urge, that of sex, is denied to the honeybees that visit flowers, because they are the workers, sterile females who perform all the tasks in the hive – feeding the queen, the drones and the young, cleaning, building, repairing, fighting off intruders – besides collecting honey and pollen. They look after the sexual needs of the flowers while having no sex life of their own. Apart from a few similarly deprived social bees, wasps and ants, this lack in their lives makes them remarkably unlike most animals, so attempts to compare their "motives," let alone their "morals," with those of others, human or not, are useless and misleading. Most modern researchers no longer try to find reasons *why* bees or other insects do things, but content themselves with reporting *what* they do.

Daumer, who had set a new standard in careful and thorough reporting with his investigations into the color vision of bees, followed up that work with an equally important analysis of the true colors of flowers as seen through the eye of a bee. By photographing no fewer than two hundred species of flowers through three different filters – the primary bee-colors ultraviolet, blue and yellow – he was able to show that the same flower may have quite a different color to a bee and to a human. Creeping Cinquefoil *(Potentilla reptans)*, for instance, which looks yellow to us, appears to the insect as bee-purple mixed with a little bee-white. On the other hand Cowslip *(Primula veris)* is yellow to bees and to ourselves because it reflects no ultraviolet. Seven quite distinct flower colors from a bee's point of view can be shown experimentally to represent no more than six to a human eye, and of those only two are the same to both insect and man.

Several mysteries have now been cleared up. Why, it used to be asked, if bees are red-blind are they attracted by some red flowers, such as the common Poppy *(Papaver rhoeas)*, from which they collect pollen in large quantities? The answer, as shown by color analysis, is that the poppy which is red to us is ultraviolet to bees. And the reason why some greenish flowers such as Stinking Hellebore *(Helleborus foetidus)* can be easily distinguished by bees from green leaves is not only that the flowers are yellow instead of green to a bee but that green foliage, because it reflects a good deal of bee-white, appears greyish to a bee.

A great many of the other pollinating insects that possess color vision at all (and some of the most primitive bugs and beetles do not, their view of the world being like an old black and white film) have been shown to have more or less the same range of perception as bees. There are, however, exceptions; some butterflies seem to be able to recognize red. It is impossible to go into all the differences here, but often what seem to us to be very subtle color distinctions will, because of the peculiarities of its visual sense, decide whether a particular insect goes for one flower or another.

To advertise itself to prospective visitors a flower may use many other visual attractions besides color in the way it dresses itself. The surface texture of petals can vary as much as that of different fabrics. Changes in the arrangement of the outer cells can give the impression of silk, velvet, wool, or even glass fiber or metallic gold and silver threads. Some petals give themselves a super-glossy surface by covering themselves with oil produced by special cells. Perhaps not surprisingly, it has been shown that such superficial adornments are much more attractive to inexperienced callers on their virgin flights than to old hands who know what they want and where to get it.

Many experiments have been made to discover the effect of the shapes of flowers in attracting insects. All types of insects have been used in such investigations, from bugs and beetles to flies, butterflies, moths and others, and once again a large proportion of the experiments have been carried out with honey-bees. Since an insect first approaching sees a flower as a two-dimensional form and the nature of that form plays an important visual role in determining whether the insect will decide to visit the flower or not, many tests have used flat paper cut-out shapes to find which is the most attractive. Sheer size is quite important, because if the piece is too small it will simply not be noticed. On the other hand, too large an area, particularly with a regular outline, is ineffective because it gives no guidance to a visitor. What attracts insects, especially the more advanced ones, seems to be contrast, and if there are no contrasts within the shape, the visitor, if it lands at all, will wander around the edges and not go towards the center, which in a real flower is the part that matters.

Experiments with increasingly elaborate shapes, progressing from simple squares and circles to stars, crosses and intricately cut shapes with "petals" like those of real flowers, have shown that the more advanced insects such as the social bees prefer long and segmented outlines; they could be said to go for the frillier shapes. This seems to accord with what is thought to have been the development of flowers from simple bowls, regular in outline and open to any visitor, no matter how crude, to the modern types, with more irregular outlines, which are only open to more intelligent visitors who know how to approach them correctly. Since these more modern flowers tend to be deeper, with their sexual organs hidden and their attractions only accessible to advanced visitors with long tongues, it is very important that the insect, which as we have seen may at first be attracted by the outline, should be stopped from merely wandering around the edge and be directed to the center, which is where the flower wants it.

Tests with three-dimensional models have shown that depth in itself is an attraction, the insect being drawn inwards by hollow trumpet or cone shapes. But many flowers have denied themselves this simple way of leading the visitor on, because they have narrowed their throat and almost closed their mouth so as to keep out unwanted intruders. To help to direct the pollinating insect towards the center of the flower, many species have developed "nectar guides," which function in much the same way as road signs. These nectar guides may consist of lines converging towards the middle, as in species of *Geranium*, or blotches leading into the throat, as in the Foxglove *(Digitalis)* or a more intensely colored zone at the entrance to the lips of the flower, as in Toadflax *(Linaria vulgaris)*. Sometimes the nectar guides may be invisible to the human eye because they make ultraviolet patterns (or because they form non-ultraviolet patterns against an ultraviolet background). Sometimes they are combined with grooves to direct the visitor's tongue to the right place. Often the visual guide is reinforced by lines or groups of cells producing patterns of scent which lead the insect on. And of course these nectar guides form the chief attraction of many of our favorite garden flowers.

As we have noted, one advertising device that some flowers use to give themselves a competitive visual advantage is sheer size. On the whole, though, that is not the most effective use of resources from the plant's point of view, because the larger the individual flowers are, the fewer of them the plant can carry; besides, a very big flower needs great structural strength to support it. The largest flower in the world is *Rafflesia arnoldii*, which measures one meter across, but it has only been able to achieve this size by being a root-parasite, which needs no structural strength because its monstrous flower rests on the ground.

An alternative way of achieving advertising impact is for the individual flowers to be small but to be so massed together that they form a large display unit between them. Such massing together of flowers, in what is known as an *inflorescence*, is the device adopted by many of the most successful flowering plants. There are two main types of inflorescence. The first is called *racemose* and consists of an arrangement where the flowers that open first are at the base of the stem and new ones open in turn towards the tip, which usually continues to elongate, forming new flower buds as it grows. The simplest in outline is a *spike*, in which the flowers are not on individual stalks; it has the advantage that pollinating insects can easily crawl from one flower to another without having to expend energy in taking off and landing. If the individual flowers are stalked, the inflorescence is a *raceme*; if the raceme is branched it becomes a *panicle*. From the advertising point of view a very showy arrangement is the *corymb*, in which the stalks *(pedicels)* of the lower flowers on the stem are long and those of the upper flowers are progressively shorter, so that the top of the inflorescence forms a sort of flat plate, which can be very conspicuous at a distance.

The second main type of inflorescence, called *cymose*, is arranged exactly the other way round. The first flower to open is at the tip of the stem; side shoots then grow out below that, each ending in a flower, followed by more side shoots below them and so on. If the side shoots grow in opposite pairs, as with Pinks and other *Dianthus*, the inflorescence is a *dichasium*;

if singly, as with the Forget-me-not *(Myosotis)*, it is a *scorpioid cyme*.

Then there is the *umbel*, in which, as the name suggests, the flower stalks radiate from a point at the top of the stem. Very often it is a *compound umbel*, as in the Carrot *(Daucus carota)*, where it is made up of a large umbel of smaller umbels.

Lastly there is the *capitulum*, which consists of a number of separate flowers closely packed together at the end of a stem so that they look like one larger flower. Examples are the Teasel *(Dipsacus)* and the Scabious *(Scabiosa)*. The capitulum has been most fully exploited by the daisy family *(Compositae)*, which is among the largest of all families of flowering plants, with some eight hundred genera and more than thirteen thousand different species. Most botanists would say that it is also the most successful family, with representatives in every part of the world where flowering plants will grow, mostly herbaceous species but including some shrubs and even a few trees. It represents a very advanced stage of evolution, and yet the typical daisy, made up though it is of a large number of small flowers, called *florets*, looks at a glance remarkably like a simple, open, single flower of the most primitive kind.

V

Sexual Taboos in Plants

Since the whole purpose of sex is to introduce variety into life, that purpose must be frustrated if any living thing fertilizes itself. The sexual equation is that: one plus one makes another; and the most important part of that equation is the word *another*. One by itself makes nothing; it can only repeat itself.

The case against *autogamy* (literally "self-marriage"), as self-fertilization is called, is therefore a very strong one. What is the use of all those millions upon millions of years of effort and experiment that it took to invent sex if any form of life properly equipped with sexual organs uses them to make itself pregnant instead of cross-breeding? The argument was put in rather moralistic terms by Raoul Francé in 1926, and shows signs of the attitude to family life in those days. Writing of the "higher" flowers, he asserted "They know how to avoid self-fertilization in a very clever way. Nature avoids incest as far as possible. She cannot always avoid it, but only in extreme cases takes refuge in it, only when no other means is available. The preservation of family life finally triumphs over scruples." Still later he extends the sentiment: "The tragedy of high aristocracy and of princely families reaches down to the little meadow flower, when it falls into the error of marrying one of its own family. It works all right for a while, but then the number of children diminishes and the existing ones become senile in their youth and get ever more and more incapable. The capacity for life dies. Nature herself raises a warning finger: being too exclusive is unhealthy."

It is true that in 1926 there were some quite outstandingly effete members of the old nobility wandering around the capital cities of Europe, living on whatever money and treasures they had managed to save from what remained of their once great palaces and stately homes. Many of them had established a rather faded social round in Vienna, where Francé lived, and he was able to formulate his somewhat puritanical view of the feebleness and decay caused by what he saw as centuries of inbreeding among a few grand families. As an observer he could hardly have been blamed for taking his examples of the bad effects of inbreeding from the chinless wonders of the decadent upper classes. It was an attitude shared by most progressive people in those years after the Great War, which had changed society beyond recovery and had, by the slaughter of millions of the strongest and healthiest young men, seriously depleted the European breeding stock for generations to come.

Francé's attitude is just another example of how the fashionable moral assumptions of the day can determine what even a supposedly objective scientist allows himself to think. Inbreeding was in fact more widespread among many local peasant communities, whose members never moved far from the place where they were born. The aristocrats did at least get around in their social wanderings, and sometimes a local girl would fall for their well-mannered charms, find herself with child and use it to force some lusty local lad to marry her. The offspring when it arrived would perhaps look remarkably unlike the legal father and set tongues wagging. It was also not absolutely unknown for the occasional Archduchess to present her Archduke husband with a son and heir bearing a marked resemblance to the stable boy at an inn where they had stayed the night on their way to a society ball.

If one is to take Francé's parallel with plant reproduction still further, it is worth noting that the aristocratic genes were likely to get more widespread among the peasantry than those of the peasants among the aristocracy. The upper classes, besides their obvious allure and greater practice in the arts of seduction, could penetrate farther in their travels because they were not tied down by the need to earn a living; they also had carriages and motor cars, and at the time Francé was writing they were beginning to use airplanes and so increase their range still farther. The peasants were, on the whole, restricted to the places they could reach on their day off on foot, or by farm cart, or by bicycle.

The result was that Francé's fears for the genetic impoverishment of the wellborn were not entirely realized; a certain amount of bastardy – or, to put it as a biologist might, outbreeding or "hybrid vigor" – brought the benefits of much-needed variety to both ends of the social scale and greatly improved the stock by enlarging and enriching the population's hereditary possibilities, or "gene pool."

There was, in any case, something basically wrong with Francé's comparison between plants and people. No aristocrat, however effete, could produce by sexual intercourse with his wife, however closely related, an exact replica of himself. The offspring would be different: not perhaps different enough to make him or her better looking, more attractive, stronger, healthier or less stupid than the parents, but different all the same. However much alike the Archduke and his Archduchess might be in looks and intelligence – or lack of it – they were not the same person, so their coupling could not really be called an act of self-fertilization. The self-pollination of a plant is quite different, and likely to lead to much closer similarities between parent and offspring. Even here, though, the offspring would only be identical with the parent if that parent – who was mother and father at the same time – was truly *homozygous* in every respect: that is, each gene on each chromosome was exactly the same as the gene at the same spot of the corresponding chromosome of the pair. The overwhelming majority of sexually produced plants are *heterozygous*, at least to some extent: that is, their genetic makeup is mixed, with many pairs of different genes. So the egg cells and sperm cells will differ from each other somewhat even on the same plant, and as a result the offspring will not be exactly the same, even when the plant is self-fertilized. (There is an exception, as in the case of human beings and other animals, when identical twins – or sometimes triplets or quadruplets – arise from an egg cell that splits up after being fertilized; but this can happen with cross-bred plants as well as with self-pollinated ones.)

There are, as we shall see in the chapter on Virgin Birth, quite a few cases where certain plants, sometimes of great economic importance, seem to get along very well by self-pollination, so perhaps there is not quite such a horror of incest in the vegetable kingdom as Francé suggested. Even with all the similarities that some people claim to have discovered between the feelings of plants and those of people, it is unlikely that plants will ever be found to be suffering from an Oedipus complex. Unfortunately, however, even eminent botanists seem to have Oedipus complexes like other people, so they share the same irrational fears of incest. They therefore tend to find "laws" on the subject in the world of nature to correspond with the manmade incest laws by which societies, from the most primitive to the so-called civilized, express their deep anxieties over the matter by forbidding certain "relations" to have sexual intercourse with each other. That these anxieties are irrational rather than genetic is evident from the fact that many of the forbidden relationships are not based on blood ties at all; as these words are being written, a young man in England has been refused permission to marry his grandfather's second wife, who is not related to him, because it is against the law.

Fortunately for the vegetable kingdom plants do not appear to have such anxieties; they indulge in inbreeding or out-breeding according to which is best for their survival and well-being, and they have no laws on the matter.

It is probable that Francé, who appears to have been something of a puritan, was also worried by a possible similarity between self-pollination and self-pollution, which was a term used at the time for masturbation. In 1926 there were still pundits – teachers, clergymen, doctors and even scientists, including botanists – asserting (in spite of the fact that they had all, unless they were seriously abnormal, done it themselves) that self-pollution was a sin that was bound to lead to awful retribution, sending those who practiced it mad, or blind, or both; if by some divine mercy they escaped such a fate they would at least become enfeebled and "prematurely senile" (almost Francé's words exactly).

It is therefore necessary not to believe too firmly the oft-repeated statement that nature disapproves of self-fertilization and will go to any lengths to avoid it. As we have seen, such statements stem from the desire to formulate universal moral principles; they make good uplifting tracts and good journalism, but they have nothing to do with real life. They represent a kind of inbreeding of thought, caused by copying without variation a principle published in 1799 by Thomas Andrew Knight that no plant can without detrimental effect pollinate itself over many generations.

Because of the fact that most of his own observations were made on the outbreeding plants that attracted the bees and other insects that so fascinated him, the great Charles Darwin tended to accept Knight's principle somewhat uncritically. With the authority of his name behind it, the principle was restated as the Darwin-Knight Law that "Nature abhors perpetual self-fertilization." This so-called "law" was, to quote Professors Faegri and van der Pijl, to play a rather unfortunate part in pollination ecology, leading less critically minded followers of Darwin to neglect the phenomenon of self-fertilization, and to search, sometimes rather frantically, for cross-pollination mechanisms and adaptations even in habitual inbreeders. Fortunately since Darwin's day scientists tend to have given up trying to fill in the gaps left by the decline of religious faith by substituting scientific laws for religious ones. They have, at any rate in the observational sciences like biology, ceased to adopt the role of lawgivers and prophets. They are content to describe *what* happens and *how* it happens, without trying to say why it happens or what the final purpose is. We can therefore forget about the Darwin-Knight law or any other law and just examine some cases of how sexual taboos operate to prevent self-mating in outbreeding plants.

Outbreeding plants are those designed for cross-pollination. They form a very large proportion of the flowering plants. In some the taboo against self-fertilization is not absolutely complete; if forced to choose between self-mating or no mating, they will in the last resort choose the former. In others the taboo is complete; self-fertilization is avoided at all costs.

Before we look at some of the ways in which the taboo operates, it is important to be clear on one point. Self-pollination does not only consist of the female organs of a flower receiving pollen from the male organs of the same flower. That, called *autogamy* ("self-marriage"), is the most obvious example of self-mating to be avoided, and for many years it was thought to be the only one. After all, one flower is the same as itself, but different flowers are – well, different. That was the view before the findings of genetics and of *cytology* (the study of cells) showed us that it was totally incorrect. Different flowers on the same plant are *not* different; the genetic make-up of their cells is exactly the same, as indeed is that of every other cell in the plant. Therefore pollination of one flower by another on the same plant would be just as much self-mating as that of male and female parts of the same flower. Though such mating between two flowers on the same plant is called *geitonogamy*, which comes from the Greek for "marriage between neighbors," it is not at all like the sexual affairs between neighbors that human beings indulge in. With a plant the two flowers are not really neighbors at all but parts of the same individual; so pairing them would also break the taboo against self-marriage.

To complicate the matter still further, the taboo must also apply to flowers on separate plants if those plants form part of the same *clone*: that is, if they have been grown from the same plant by cuttings, or by runners, or by splitting up a clump of herbaceous plant or a cluster of bulbs. They are all identical in genetic makeup, so that pollinating one by another from the same clone is still self-marriage.

With such problems to contend with to avoid breaking the taboo, different species have used an enormous variety of different methods to overcome the difficulties. The question may be asked why plants did not take the easy way out by doing what human beings and most animals do: having each individual either male or female, but not both. Then there would be no possibility of any plant fertilizing itself, because it would not have both the male and the female equipment with which to do so. Some species do have their male and female flowers on separate plants. These are called *dioecious*, from Greek words meaning "having two separate homes." In certain cases the arrangement seems to work very well; but, as might be expected, for the male and the female to occupy separate homes does tend to be rather expensive. The stinging-nettle does it, and so does the holly (though sometimes an odd holly tree will turn out to be polygamous). This keeping apart of the sexes is thought to be a recently acquired habit by a few plants that used to have the sexes together. One of the reasons for this belief is that in many dioecious plants there is clear evidence that the flowers have not been single-sexed for very long. A well-known case is that of the Red Campion *(Silene dioica)*, whose male flowers contain rudimentary female sex organs and whose female flowers have half-formed male parts in them, which strongly suggests that they have only recently given up being hermaphrodite.

The reasons for the abandoning of the hermaphrodite state can only be guessed at, but no doubt will usually have been adaptation to some change of circumstances. However, one-sexed flowers are the exception. Most normal flowering plants are hermaphrodite, in spite of the difficulties this creates for them in trying to avoid self-fertilization. The reason why it is expensive to have the sexes on separate individuals, as men and women do, is that it puts plants at an economic disadvantage. The trouble is that males cannot become mothers. So if half the population is male and half female, then the half that is male is quite wasted after it has performed its function of getting the other half pregnant. It is merely occupying space that could be more profitably used by plants capable of having young. An expectant human father is a fairly useless object when the woman he made pregnant is about to have a baby, but at least he can provide support for her and food and protection for the family. A male plant can do none of these things; after it has provided the only thing it can contribute, sperm, it has no further function or responsibility.

Hermaphrodite plants have double the chances of producing more offspring, and so occupying ground that would otherwise be seized by more fertile competitors, because each plant can bear young; so it is hardly surprising that the hermaphrodites far outnumber the species with only one sex per plant. But it brings the problem of how to escape self-fertilization, and different species cope with the problem in different ways.

The most obvious way is to have the whole plant hermaphrodite but to have separate flowers on it, some male and some female, but none of them both. Species of this type are called *monoecious*, from the Greek words meaning "single home," because although the male and female flowers are separate they live together on the same plant. To use human legal parlance, they cohabit. How can they be made to do so without also copulating? The answer is not quite so simple as might at first appear. Not only is there the likelihood that the slightest shaking of the plant, by a breeze or a passing animal, will cause the pollen from a male flower to fall on the stigma of a female flower, but a visiting insect may fly or crawl from one flower to another and so bring about an unwanted self-pregnancy.

There are some monoecious plants that are pollinated by insects, such as cucumbers and squashes, and flowers of these have to be sufficiently large and attractive to persuade the insects to visit them. They offer nectar and pollen as a reward, and seem not to be too bothered about self-fertilization, since bees can often be seen going from male to female flowers and vice versa on the same plant. Flowers in this family tend to attract large, furry bumblebees, whose long hairs are perfect for entangling the rather heavy pollen which they carry on their bodies from flower to flower. The fruits of these plants are often placed behind the female flowers instead of within the flower as in most cases; it has been suggested that the reason for this is to protect the fruit in its delicate early stages from the damage it might suffer by rough treatment from the coarse mouthparts of the heavy, rapacious bumblebees when they were foraging inside the flower. Perhaps the apparent abandonment of defenses against self-fertilization in these species might not be as complete as it seems; there may be taboos working at the deeper levels of pollen preference and incompatibility, as we shall see later.

Many monoecious species make no attempt to invite insects,

and rely on the wind to carry the pollen from the male to the female flowers. One such species is the maize, which as we have seen was used a great deal in the early experiments. These, paradoxically enough, used castration both of the male and of the female organs to demonstrate the existence of sex in plants because there were such large and obvious differences between the flowers of opposite sex, and no skill or delicacy was needed to cut off the male or female parts. With no necessity to dress themselves up with colored petals or any other glamorous inducements to attract the insects, the separate male and female flowers could strip themselves of all clothing and become naked sexual organs, with no other function than on the one hand to produce pollen and on the other hand to get fertilized.

That is why on so many of these plants with wind-pollinated single-sex blossoms the flowers are so very retiring and unobtrusive; the wind needs no show to make it blow. Once again, with these separate male and female flowers reduced to naked sex organs for the wind to play with as with the brightly petalled bee-flowers of the squash, self-fertilization is probably very common. Still, if by processes within the female organs the sex cells from the pollen of other plants are given a better welcome than those from pollen from the same plant, at least some of the resulting offspring will be outbred instead of inbred. And if the offspring from the mixed mating (called *xenogamy*, meaning "marriage to a foreigner") are stronger and healthier than the inbred ones, they will win the competition for survival and so increase their proportion in the plant population.

Most of the monoecious plants do not carry their male flowers as separate units but in long inflorescences, which swing much more widely in the wind than single flowers could, and so scatter their pollen over a bigger area, where at least some of it has a chance to reach other plants. The female flowers, on the other hand, are usually produced either singly or in small clusters of no more than two or three, because they are the ones that will produce the fruit, and so need room to grow and swell. The string of male flowers, often called the catkin, is of no further use after it has emitted its pollen, and soon it withers and falls off.

A typical species of this kind is the hazel, with its long,

dangling male catkins swinging in the spring breeze sending golden clouds of pollen into the air, and so being one of the first plants every year to start the hay fever season for those allergic to pollen. Each of the millions of pollen grains which form the cloud is the potential father of a nut, and so of a new hazel tree. But first it must fall on the small tuft of red hair projecting from one of the half-hidden female flowers. On that tuft of hair, which is the beginning of the female sexual organs, the style, there are perhaps hundreds of other pollen grains. The grain that will become the male parent of a new nut is the one that most successfully makes its way through that tuft of red hair by developing a projection, the pollen tube, which grows and lengthens, penetrating through the female tissue till it reaches the ovule and releases the male germ cell which will unite with it to form the new individual by fusion of the male and female cells. The other pollen grains will be sending out competitive pollen tubes, each forcing its way through the female tissue in an attempt to win the race and become the father instead.

There are many things that decide which will be the winner. One is which grain was lucky enough to fall on that tuft of red hair first; that will put it ahead of the rest. On the other hand, when it comes to penetrating the female tissues with their tubes, some get started quicker than others, and the first one on the spot may be slow to get moving and be left behind. But the ones that start quickest do not necessarily have the staying power; they may penetrate a certain way and then give up. But even those that go the whole way differ in their performance. The winner in the end is not just the one that is quicker, or pushier, or longer lasting, but the one that is most attractive to the female part of the encounter. All the time that the process has been going on, hormones have been secreted and rapidly diffused into the tissues, encouraging some of the contenders and discouraging others. In a very real sense the female organs are in control. It is true that the winning pollen has to be competitively superior to get to the ovule to fertilize it, but that is not enough. If the genetic makeup of the male is incompatible, the female will sense it chemically and by chemical countermeasures put up barriers to its approach; if even then the male germ cell gets to the ovule and tries to fuse with it, real incompatibility will make sure that either the fusion does not take place or, if it does, that no seed is formed or that the seed is infertile.

Since in a large number of species the female organs prefer "foreign" pollen to pollen from the same plant or the same clone as themselves, this mechanism for rejecting the incompatible may in the last resort give protection from self-fertilization. However, there is a snag. If there is one hundred percent total incompatibility between the male and female cells of the same plant or clone, then there will be no possibility of breeding any offspring at all by self-fertilization if all else fails and there is no pollen from another source around to effect cross-pollination.

There seems to be little doubt among botanists at the present time that among the higher plants separate male and female flowers, whether on the same or on different plants, do not as once thought represent more primitive forms in evolutionary terms but are later developments from hermaphrodite flowers, and some would say represent not improved but degenerate forms. There are, as we have seen, some successful examples of these plants with single-sex flowers, but they may have reached the end of an evolutionary line and have no potential for further development. This is for various reasons, including the one we have already touched upon, namely that it is a great waste of resources for half the flowers in one case and half the plants in the other to be unable to produce young.

There are some odd intermediate forms between plants with wholly hermaphrodite flowers and those with separate male and female flowers. Any combination that can be thought up, no matter how odd, is almost certain to have been tried by nature as an experiment during its "cosmic doodling." Some species, called *androdioecious*, have only male flowers on some plants and only hermaphrodite flowers on others. Some, called *andromonoecious*, have separate male and hermaphrodite flowers on the same plant. Others, called *gynodioecious*, have only female flowers on some plants and only hermaphrodite flowers on others. Then there are those called *gynomonoecious*, which have separate female and hermaphodite flowers on the same plant. And lastly, to exhaust every sexual possibility there are those with male, female and hermaphrodite flowers, either on the same or on different plants; the botanists, rather misleadingly, call them polygamous. Mostly these in-between plants seem happy to fertilize themselves, but manage to have enough outbreeding, aided by a certain amount of incompatibility with their own pollen, to ensure that there is enough "new blood" constantly being brought in to avoid deterioration of the stock.

The normal thing, however, is as we have seen for flowers to be bisexual, and capable of performing both the male and the female function. That is the state of the majority of plants, and that is where they face their greatest sexual problem: how to avoid, or at least to discourage, self-fertilization when the flowers are hermaphrodite so that the male and female organs are close to each other all the time.

There are basically two ways in which self-pollination can be avoided in these hermaphrodite flowers. They both involve the separation of the male from the female, so that they are prevented from coming together, either directly or through the agency – one might say the matrimonial agency – of a third party, usually an insect but sometimes some other creature, which during its visit to the flower transfers male pollen to female stigma, and so brings about sexual contact as effectively as if the male and female organs had been allowed to touch each other directly. The first of these methods of separation of the sexes is in space and the second in time.

Physical separation in space, called *hercogamy*, which means barrier to marriage, is the common condition of these bisexual flowers, in the sense that it means that the male and female organs are separated from each other. The term is also often used for cases in which visitors meet with some structural barrier or mechanism which prevents them from carrying the pollen to the stigma.

A famous case of such a structural mechanism is that of the common primrose, which was observed and commented on by Darwin in the painstaking detail for which he was renowned. The mechanism involved is called *heterostyly*, and it works like this. Though all primrose flowers may look alike at a casual glance, there are really two quite different types: pin-eyed and thrum-eyed. In the pin-eyed flowers the stigma is visible, like the head of a pin, sticking slightly up at the center of the flower. That is because the style, which carries the stigma, is long, and the stamens, which carry the pollen, are hidden inside the flower, where they are fixed halfway up the floral tube; if you cut a flower in half lengthwise you can see them there in the middle. In the thrum-eyed flowers, on the other hand, it is the stigma that is hidden inside the flower and the stamens that are visible at the top of the flower. That is because in this case the style is short, so that the pinhead-like stigma only comes halfway up the tube, to the same height as the stamens in the other type of flower. Pin-eyed and thrum-eyed flowers are always borne on different plants, and there are just about the same numbers of each in any wild population of primroses. In 1938, the scientist J. B. S. Haldane, who was so keenly interested in genetics that he put his ideas on the subject into practice by breeding a new strain of cat, published the results of a survey he had made by patiently counting no fewer than two thousand three hundred and two primrose plants in different places and seeing how many of each there were. He found an almost identical number of plants of each type.

In addition to the more obvious distinction between the length of style and position of stamens in the two kinds of flower, there are other differences. The pollen grains on the thrum-eyed flowers are bigger, and so fit better into the slightly coarser surface of the stigma of the pin-eyed flowers; the smaller pollen grains from the pin-eyed flowers, on the other hand, fit more neatly into the rather more finely textured surface of the stigma inside the thrum-eyed flowers. The arrangement is admirably designed to make sure of a high degree of cross-pollination between the two types of flowers. Darwin carefully examined the different results obtained by cross-pollination and by the pollination of each type of flower by one of its own kind, both from the same and from a different plant. His findings supported his view that crossing is much superior to self-fertilization, both in numbers of seed produced and in the vigor of the resulting seedlings. What were called, rather moralistically, the results of "illegitimate" union between like flowers compared very unfavorably with the results of "legitimate" cross-breeding.

Far fewer seeds were set from "illegitimate" than from "legitimate" mating. Under natural conditions, it was later realized, there would have been fewer still from the "illegitimate" fertilization, because Darwin's experiments had artificially favored the "illegitimate" pollen by giving it no competition. In further experiments, where a mixture of "legitimate" and "illegitimate" pollen was used, it was found that secretions from the female organs were able to choose the "legitimate" pollen by speeding up its penetration of their tissues with its pollen tube. One experiment consisted of putting "illegitimate" pollen on the stigma, and then waiting twenty-four hours before applying "legitimate" pollen as well. In spite of its time handicap the "legitimate" pollen reached its goal first and succeeded in bringing about fertilization, because the female parts had welcomed its advances in preference to those of its rival.

Only now do we know the genetical facts, which there was neither the knowledge nor the technique to find out in Darwin's day. The breeding system of the primrose rests on something he could not have known: that the plants with pin-eyed flowers have the same gene twice for this character (they are *homozygous recessives*), but those with thrum-eyed flowers have a pair of different genes (they are *heterozygous* for style-length). Because of this, cross-breeding is strongly favored and, as may be predicted from genetic theory, equal numbers of each type are produced as the result.

The beauty of the simple mechanism of heterostyly is that it makes crossing almost certain, whatever type of insect may visit the flowers. The primrose caters to different visitors with different lengths of tongues. Bees and butterflies have been seen at the flowers, and sometimes those quick, darting mimics the bee-flies, and Darwin thought that night-flying moths might also be important pollinators; each of these creatures will visit both types of flower and give them the same shallow or penetrating treatment with its tongue. With such efficient devices for cross-fertilization and such a range of visitors, coupled with a limited possibility of self-fulfillment if no visitor turns up, the primrose's future seems assured. Or rather it would be if the species could develop some equally effective device for protecting itself against its most ruthless enemy, the human beings absurdly called "flower lovers" who pick its blossoms every year and who are almost making it extinct in some places.

Many other *Primula* species, such as the Cowslip, the Oxlip and the Bird's-eyed Primrose, show a similar mechanism of heterostyly to enforce the taboo against self-fertilization. The efficiency of the mechanism is shown by the fact that it has been developed by members of quite different families as well, such as the pretty *Pulmonaria*, called Lungwort because of a fancied resemblance between its white-splashed leaves and the human lung, with its two forms of flower both starting pink and then turning a shade of blue that rather clashes with the pink like an ill-designed color scheme.

The mechanism is taken even further by some other species, whose breeding system rests not on two but on three different forms of flower found on separate plants. One such is *Lythrum salicaria*, the Purple Loosestrife, of which cultivated varieties are grown in garden borders, where they are kept going from cuttings or the splitting up of overgrown clumps, and where in most cases they have no hope of a proper sex life, because only one form is grown with only one kind of flower. In such cases the taboo against self-pollination is repressive enough to make sure that the plant either has no sexual fulfillment at all or one or two unsatisfactory experiences. Out in its natural surroundings the Purple Loosestrife has one kind of flower with a short style and long and medium stamens, a second with a medium style and long and short stamens, and a third kind with a long style and short and medium stamens. Each form is found on a different plant. To add to the complexity, three sizes of pollen grain are found, one in the short stamens, one in the medium stamens and one in the long stamens.

The same device of three types of flower is found in other plant families, such as those of the Wood Sorrel, the Flax, and that exquisitely beautiful menace of the world's tropical waterways the Water Hyacinth. That the mechanism works there can be no doubt, but many botanists think that separation into different types of flower like this may have led in the end to single-sex flowers on different plants.

Many other examples of different forms of flower within the same species could be given, but we will make do with one last example that is very simple, very neat and very effective. In this case it is not the flower as a whole that is different in different plants, or even that the sexual organs are put in different positions, but that the difference is confined merely to the pollen and the stigma. The plant is the Thrift or Sea Pink, *Armeria maritima*. At first sight, the flowers all look the same. But closer examination by means of a magnifying glass – or the naked eye for those with good sight – will show that there are two kinds of pollen and two kinds of stigma. Some plants carry flowers with coarsely surfaced pollen and a delicately sculptured surface to the stigma; other plants carry flowers with finely surfaced pollen and a coarsely surfaced stigma. It is a beautifully simple way of ensuring cross-pollination. The coarse male pollen from one type of flower hits it off with the coarse female stigma of the other, and the refined pollen from the second flower finds that it was made for the refined stigma of the first.

It may be asked why, in those cases where there is strong incompatibility between a plant and its own pollen, it should be necessary to get up to all these dodges to see that it does not get self-pollinated. The answer may be a very simple one. A flower is open only for a limited period, and the female stigma may not be at its best for very long. If therefore unsuitable pollen, which will either not be able to mate at all or will do so with poor results, should fall on the stigma it will be wasting that stigma's time and perhaps crowding out a more suitable contender. It seems sensible for there to be a mechanism like an old-fashioned parent which says "Don't waste your time with unsuitable males; get hold of the right one before you are faded and past it."

So much for avoiding self-mating by separating the male and female parts in space. The other method, which has much to recommend it, is *dichogamy*, or separation in time. It relies upon the ripening of the male and female parts of the flower at different times, so that when the stigma is ready for sexual intercourse the pollen is not, and the other way round.

In most cases – as with human beings, to judge from the letters in sex problem magazines – the male is ready before the female. A good example of this is the Rosebay Willowherb, *Chamaenerion angustifolium*, often known as Fireweed because it rapidly appears, as if from nowhere, on the sites of buildings that have been burned down and on waste ground, which it decks with its tall spikes of bright rose-purple flowers and later its plumed seeds, with which it extends its territory, as if to cover up the ruins as quickly as possible. Fifty years ago it was rather a shy local plant, but the spread of urban decay has increased its chances greatly and made it the dominant flower in many run-down areas of cities. The stamens grow out from the clawed petals and shed their pollen while the female parts are small and immature. Not until the pollen has all been shed and the anthers have withered does the style reach puberty and thrust itself out beyond the stamens, becoming sticky and receptive to catch the grains from another plant still in its male phase.

There are many examples of this *protandry*, as this readiness of the male parts before the female ones is called. One of the most striking is the Yellow Mountain Saxifrage, *Saxifraga aizoides*. To make absolutely sure that it cannot be fertilized with its own pollen, it castrates itself after its male phase is over, cutting off and dropping its anthers as soon as they have finished emitting pollen. At the time this self-castration occurs the female part of the flower is quite undeveloped; all that can be seen is a large, disk-shaped nectary gleaming with the nectar which it produces in great quantities for the delight of the many flies it attracts. Not till some time has passed after the castration does the flower develop its female personality, so that any pollen left lying around will be too old to be any use. After that, the female parts grow rapidly, swelling out and developing nipple-like stigmas which become receptive to pollen from other, still uncastrated flowers.

There is a great finality about castration, though. It completely rules out the possibility of begetting any more offspring, so if in spite of everything cross-pollination does not take place there is no chance of self-fertilization in the last resort; that may be why some flowers of *Saxifraga aizoides* cheat a bit on occasions, by leaving the odd anther on when the rest have dropped off.

In passing, it may be noted that female as well as male castration is practiced in the plant world. Where the female organ is only sexually responsive for a short time, it may be an effective way of avoiding unsuitable contacts and preserving desirable cross-mating. The Shoo-fly Plant, a beautiful blue-flowered member of the potato family from Peru, is a striking example. As soon as the plump stigma has been pollinated it starts to wither; within an hour it has completely shrivelled, and the whole style drops off. This self-castration, swiftly arranged by reactions from the tissue of the style as the pollen tube penetrates it, removes all the flower's exernal female organs in one operation, leaving it incapable of further sexual response so that it can confine itself to the job of producing offspring.

Usually something rather less drastic than castration is involved. Take the case of Rue, the "herb of grace" mentioned in Shakespeare. The flowers of this little shrub, *Ruta graveolens*, stop short of self-emasculation. Soon after the flower opens, the stamens pull themselves out of the pocket-like petals and bend upwards till they stand erect. When the anthers open and emit their pollen, the female part is too young for sex; the style is very short and the stigma undeveloped. During most of the time that the pollen is being shed, the female part remains under age, but the style starts to grow and the stigma gets ready to perform its function. By the time the style is fully grown and the stigma has become sexually receptive, the stamens will have exhausted themselves and lost their erections. If all goes well, that stigma will receive pollen carried by an insect from another flower and cross-pollination will have been achieved. If not, self-pollination will be brought about by a last effort on the part of the stamens, which will bend themselves up again and become erect enough to touch the stigma and transfer a few last pollen grains to it.

A similar gallant last-ditch male effort is recorded of the well-known decorative climber *Cobaea scandens*, which in its native habitat in South America is pollinated by bats. If a bat fails to turn up, or carries no pollen, or the pollen it carries is unsuitable, so that cross-pollination does not take place, a last movement of a stamen in the unsatisfied flower will bring about self-pollination.

Many species of *Campanula*, the Bellflower, show a rather more precise and organized approach to the same problem: how to ensure as far as possible that its flowers are not self-fertilized, but to leave itself that option if all else fails. Most of them are visited by rather large insects such as bumblebees; the visitors have to be big and strong to penetrate with their tongues to the back of the flower in order to get at the nectar produced by a ring of special tissue at the base of the style. The anthers, which form a ring around the style, open inwards and deposit their pollen on the middle of the style, which catches and holds the pollen grains in its thick covering of hair. It may seem as if this is sure to bring about self-pollination, but at that stage there is no danger of that happening, because the surface of the style is quite unable to respond sexually and the stigma at the end of the style is tightly closed. The next development is that the style starts to elongate. As it stretches, the hairy part is pushed forward, with the pollen grains still stuck to it; but the stigma is still tightly shut. At the next stage in the flower's sexual history, the male organs, having fulfilled their function, shrivel up completely, so there is no hope of any further sexual contribution from them. The stigma at last begins to open out, rather cautiously, into three lobes, ready to receive the pollen which, if all goes according to plan, will be deposited on them from another flower by the next bee that comes along. The device is usually successful, as can be seen from the flourishing state of most *Campanula* species. If cross-pollination does not occur, however, the flower has an absolutely certain way of pollinating itself which in no way relies on chance as in the previous examples. The lobes of the stigma open wider. As they do so they start to roll back, and they continue to do so till at last they come in contact with the pollen-grains which have been sticking to the hairy style ever since they were put there by the anthers of that same flower when it first opened. The favorable conditions of protection and enough but not too much humidity inside the flower will have kept those pollen grains fresh and capable of functioning.

The device of protandry, or starting as a male, is so widespread in the plant world, and has so many variations, that the rest of this book could be devoted to this subject alone. We will finish with an example that has a certain narrative drama about it like a somewhat implausible cliffhanger. It concerns one of the Milkwort family, *Polygala chamaebuxus*, a native of the European Alps. To attract pollinating bees, the flowers have large upper sepals of a luminous purple-mauve color; the petals form a purple tube and the outer parts are yellow. The color scheme is just right for bees, which regularly visit the flowers, landing on the lower petal, which is hinged and is pushed down by the weight of the bee. In this case the protandry is so marked that the anthers open and emit their pollen while the flower bud is still closed. This pollen is deposited on a sort of platform which is fixed to the style. As the flower bud opens the style starts to lengthen, and it continues to do so till the platform with the pollen on it is carried up like an elevator, leaving the anthers behind.

At the base of the style, around the ovary, nectar is produced which is extremely attractive to bees. But the nectar can only be reached through the tube formed by the upper petals, which opens backwards. To complicate matters, there is a pocket in each petal which forms a blind alley; it is thought that the delay to the bee's progress caused by these blind alleys may cause it to wander around a little longer in the flower and so help pollination. When it has found the right direction, the bee pushes on through the tube to get at the nectar. But the tube, being formed only by the top two petals, has no floor. The bee, losing its footing, drops on to the lower petal and causes the hinge to operate, so that the petal is lowered abruptly. That causes the style to emerge and give the bee a sharp bang, covering it with pollen from the platform. That pollen will be carried on the bee when it visits the next flower, where it will be deposited on the stigma just before the trap door mechanism operates once more, covering the visitor with fresh pollen.

Sometimes the female is ready for sex before the male. This method of avoiding self-fertilization in hermaphrodite flowers is known as *protogyny*, meaning "woman first." It is well known in quite a number of plant families, though by no means as common as protandry. This is not surprising, because it reverses the usual sex roles. The "normal" sequence of events in a flowering plant as it develops is that the calyx starts to form before the petals, the petals before the sexual organs, and the male organs before the female. When the female comes before the male, therefore, it could be said to be "against nature." To justify such a reversal, there would have to be some very real advantage in the cases where protogyny occurs. And there is, in the opinion of many botanists. The argument goes like this: if the female organs of a flower are ready before the male ones, then pollen from other sources will have a head start and so be able to carry out fertilization before any of that flower's own pollen appears. It might then be argued that these "female-first" flowers are more advanced than "male-first" ones and represent a higher form in the evolutionary scale. And that, in many cases, seems to be true. There is, however, no simple evolutionary path from lower to higher. In the development of new forms, whether in plants or in other forms of life, there is no simple straight line always leading from lower to higher. Some species seem to have gone up, some down and some sideways. So those who would see in the development of protogyny among plants a line of advancement similar to the women's liberation movement among humans may be taking too simple a view.

A common example of a plant whose flowers begin as female and then change to male is the Figwort *(Scrophularia nodosa)*. The color of the flowers is the sort of reddish-brown that seems to appeal to wasps, which visit the blossoms eagerly in search of nectar. For the first two days after the flower opens it is almost aggressively female, with the anthers bent double on the stamens, out of the way of visitors, and the style protruding from the mouth of the bell-shaped blossom so that its receptive female stigma cannot fail to press itself against a visiting wasp. Having in this way removed from the insect any pollen it happens to have on it from a previous encounter, the flower rewards it with nectar produced by a ring of tissue at the base of the female organs. When its two days as a female are up, the flower changes sex; the style bends down so that the stigma no longer touches visitors, the stamens straighten out and become erect, exposing their pollen-laden anthers, and the flower becomes thoroughly male.

The timing mechanism which tells a flower when to change from one sex to another is imperfectly understood. Species of a close relative of the figwort, commonly called Musk or Monkeyflower *(Mimulus)*, have a very remarkable timing device of a different kind. As soon as the stigma is touched, its two lobes close tightly together; if it is a compatible pollen that they are gripping they remain closed, but if not they open again after a time to give themselves another chance.

In a large number of examples of plants in which it could be said that the female takes the initiative, there is some kind of trap involved. The case of the California Allspice, *Calycanthus occidentalis*, is a fairly typical one, though the trap is perhaps a rather gentle one compared with some others. The flowers of the California Allspice, an attractive shrub with aromatic leaves, have a large number of petals (technically tepals) which are purplish-red tinged with brown. When the flowers open, giving off a strong scent like a rather fruity wine, the outer parts fold back, but the inner ones make themselves into a cone, hollow inside. For the first day or two after opening, the flowers are at the female stage. A small beetle, drawn by the smell, crawls into the dome-shaped chamber, where it finds a number of granular white food bodies, looking rather like semolina, on which it feasts. When it has had enough, it tries to get away, but downward-pointing bristles stop it from getting out, and it remains a prisoner, together with any other beetles which joined it for the feast. (It has been suggested that the beetles were not prisoners but liked it there, with plenty of food and a lovely smell; but somebody proved otherwise by cutting a flower open, whereupon the beetles escaped as fast as they could; perhaps the smell had become too much for them after being confined with it, and each other, for so long.)

In a day or two the flower changes over to the male state. The stigmas, which have been protruding into the chamber since the flower opened, quickly wither. The anthers on the stamens open and powder the beetles over with pollen; to make sure that the pollen does not get onto any stigmas still unwithered, stamen-like growths without anthers have grown to make a protective covering. When the beetles have received their powdering with pollen, the top of the cone opens, the tepals fold back, and the beetles are set free; whereupon they go straight to another flower just open and at the female stage, carry the pollen from the first flower to the stigmas, have a feed, find themselves trapped, and face another day or two in prison.

Many protogynous flowers use some form of trap, but some offer no food or anything except a luring smell, some sentence their visitors to longer terms of imprisonment, or even to death, and some have developed devices of quite refined cruelty. Maybe, to alter Kipling's phrase, the protogynous species is more deadly than the protandrous.

To finish this chapter, let us examine two rather unusual cases of flowers with different sexes at different times. In trees of the Avocado, *Persea americana*, the flowers actually open twice, once as males and once as females. When they open first, only the female organs are ready; when they open the second time it is only the male organs that function. What is more, there are two types of trees, of which one has its flowers male in the morning and female in the afternoon, while the other type has its flowers female in the morning and male in the afternoon.

The other example sounds rather more farfetched but is quite true. A kind of Jujube, *Zizyphus spina-christi*, also has two types of trees. In this case, however, one has flowers that are male in the morning and female in the early afternoon, while the other has flowers that are male at noon and female in the night and in the morning. There seem to be few lengths to which some plants will not go in order to avoid fertilizing themselves. The ones that are best adjusted, however, seem to be those which if all else fails are capable of giving themselves sexual satisfaction.

VI

Sexual Techniques

Every living thing that wants sexual satisfaction has to offer something in exchange. It may be real or it may be imaginary, but the promise it holds out must be exciting enough to attract the right sort of attention from the right sort of visitor and to persuade that visitor to perform the necessary function.

As we have seen in the case of the Bee Orchid and other similar species, the lure may be entirely illusory and offer the visitor a sexual satisfaction that will never be attained. Sexual deceit is, however, only one of many forms of deceit practiced by plants upon their victims. We will look at some examples later, merely noting here that the more sophisticated forms of deceit demand a certain amount of intelligence on the part of the deceived. As with human beings, the victim of a confidence trick has to possess smartness beyond a certain level in order to be fooled; real stupidity offers some measure of protection, because stupid creatures simply do not grasp what bogus proposition is being made and so do not fall for it.

Before we go further into the subject of deceit, however, let us examine the genuine satisfactions that plants are able to offer their patrons. We have already seen that advertising plays a vital part in drawing attention to what is being offered; desire has to be aroused before it can be satisfied. Now we will look at the nature of the goods and services available to pollinators once their attention has been gained.

There is no doubt that by far the most important urge that a flower can exploit to its advantage is the urge of a visitor to find food. A great many pollinating insects are only interested in the immediate satisfaction of their own hunger or greed. This is particularly true of the more primitive creatures, which have no further parental responsibilities when mating is over and the eggs have been laid; once the young have hatched out they are on their own and must feed themselves. Some of the more advanced visitors, such as the social bees, collect food to feed the brood; they are of the highest importance as pollinators, since members of their working class spend their whole adult lives industriously laboring among the flowers until they wear themselves out.

Pollen itself is the most obvious source of food which the flower can offer to visitors. There is general agreement among biologists that it was in fact the first, and probably for millions of years the only, thing that attracted insects to flowers. Indeed it is considered certain that before there were such things as flowers at all, or even pollen, insects had long been eating the spores by which the more primitive plants reproduced themselves. As we have seen, pollen is a highly developed form of microspore, and it is natural that the beetles and their kind which had been feeding on spores of the earlier plants (and still do on those of their present-day representatives such as the tree-ferns) should continue to do so with the pollen of the seed plants. In the case of the gymnosperms such pollen eating could hardly be considered a necessary or useful part of pollination, even though no doubt a certain amount of pollen was accidentally transferred from the male to the female organs in the process. Since the gymnosperms were adequately pollinated by the wind – and still are, as can be seen from the flourishing state of such living examples as the conifers – their attractiveness to pollen eaters could be called self-destructive, the only ones to benefit being the beetles and other insects.

However, the relationship with insects on which the development of flowering plants has depended is thought to have originated from that early one-sided association between the eater and the eaten, the robber and the robbed, and progressed to a condition of mutual benefit and dependence, even to a point where the plant gains everything and the insect nothing, as in the case of non-orgasmic pseudocopulation.

Pollen is an excellent source of food, especially rich in protein. Its analysis reads like the formula of an ideal diet for someone watching the calories: up to thirty percent protein, between one and seven percent starch, from one to ten percent sugars, five percent fat on average, and up to ten percent ash. The outer coat, the *exine*, is very tough and resistant; that is why pollen grains are almost indestructible and can be found intact in the fossil record after many millions of years, enabling paleobotanists to tell with great accuracy the nature of the vegetation of ancient times. The surface of the minute grains can be seen under the microscope to be sculptured into patterns of great delicacy and complexity, each pattern being unique to a particular species; so precisely can plants be identified from these patterns that scientists have been able to solve many crimes such as murder and rape by examining pollen grains found on the body and clothing of suspects. Such surface markings, which look like incredibly fine ornamentation by a skilled engraver, can, as we have seen in the case of Thrift *(Armeria maritima)*, ensure that the right pollen fits the right stigma in the manner of a matching lock and key. The roughness that the pattern gives to the surface also helps the pollen grains to cling to the body of a visiting insect until it reaches another flower instead of falling off and being wasted. In addition the grains often have a sticky surface which makes them adhere to any creature that brushes against them.

Because of the extremely resistant exine as outer coat and a tough inner coat, the *intine*, made up largely of cellulose, pollen grains are not easily chewed except by beetles and other primitive insects with strong and rather coarse mouth parts; and it was, as we have seen, just such creatures which were the first to be attracted by pollen. However, the outer part of the grains is too tough to be chewed in the same way by the larvae of bees, and yet pollen forms a vital ingredient in their diet, providing the protein they need in order to grow. How can these larvae get the nourishment from pollen grains collected for them by the workers if the wall is so tough and indigestible? Examination of the digestive tract of the larvae has shown that the outer shells of the grains remain more or less intact but that the contents have gone; it seems that they are extracted through the tiny holes and thin places in the wall created by the sculptured pattern.

It is suggested that when insects go foraging for pollen they may be first attracted by the smell, either of the grains or of the sticky or oily substances covering them, which botanists have identified in many cases as a "spermatic odor." Since pollen does contain the male reproductive substance, it would seem likely that it should emit such a smell, though most people are only aware of it in certain specially odoriferous species. Probably it was that smell which first attracted the beetles by playing on some kind of primitive sexual urge, and it was only after their olfactory senses had led them to the pollen that they tried eating it and found that it was good. It is also suggested that bees and other insects fed on pollen during their larval days are strongly impelled to gather pollen when they grow up and start work because of a memory of the taste and smell of the food they had in infancy, just as we can be vividly reminded of our childhood by a sudden whiff of some almost forgotten scent.

So much attention is paid to the nectar-gathering activities of bees in order to make honey that it is sometimes forgotten that pollen collecting for the young is usually a great deal more important. There are very many different ways in which bees, the chief collectors of pollen, manage to transport it back to the hive or nest. Some of them simply swallow it, carry it home mixed with nectar in their crops and then regurgitate it. Most kinds of bees carry it on the outside of their bodies between hairs specially adapted for the purpose. Those that collect pollen with their abdomens have their undersides densely covered with hairs which curve towards their tails. The hairs are of different kinds in different types of bees, in some being unbranched and in others branched in various ways. The pollen grains are gathered up and transferred by means of the legs to the bush of abdominal hair. For this purpose the legs have special stiff brushes and combs with which the grains can be swept into place. Some bees collect pollen by letting their legs hang down and scraping up the grains as they visit flowers; they pass the pollen back to their hind legs, which they then lift up to their bodies, and in this way they transfer the grains to their abdominal hair.

Other bees transport the pollen home on their legs, which are equipped with special hairs designed to collect and carry the grains. There are very many different arrangements for the purpose, with branched and unbranched hairs, brushes, and in some cases baskets formed by dense fringes of hair. A cover to the pollen-carrying basket may be provided in the shape of specially long hairs which prevent the grains from spilling. Some bees neatly pack the pollen from one flower into their receptacles before moving on to the next, brushing it into place by a quick and efficient sequence of leg movements. Others save time by packing the pollen with their legs while they are flying from flower to flower. There are some bees that also make the grains easier to carry and less likely to be spilled by moistening them with regurgitated nectar.

The most remarkable and highly developed apparatus for collecting and transporting pollen is that of the hive bees and some other social bees. The outside of one of the joints of each of their third pair of legs carries a much enlarged and developed pollen basket called a *corbicula*. By using rakes, combs and brushes, by moistening the grains with honey and with the help of a structure called the pollen press, these corbiculae may be filled till they bulge with a compact mass of sticky pollen plastered together. When full, a special bristle on the leg is pinned through the pollen mass to hold it in place. The number of pollen grains carried in a fully laden corbicula on a bee's leg can be as much as a million, making a considerable weight for the bee to carry back home.

In providing pollen for insects to take away as food, a flower is giving away some of what might be the future of the species, so from the plant's point of view there must be at least enough surplus grains carried inadvertently by the insect and deposited on receptive female organs to make up for the amount lost. All taking of pollen could be called theft, but the term thief is usually applied only to those insects which by reason of size or habit never transfer grains to a stigma; those which carry out pollination, however unwittingly, are said to have earned the pollen they took as food.

A typical pollen flower exposes its sexual organs to all comers in what biologists call a promiscuous way; it consists of an open, bowl-shaped blossom, usually of a fairly primitive type with separate petals, and its visitors need no period of getting to know it and learning its ways. One example is the common Field Poppy *(Papaver rhoeas)*. Its flowers last only a day, so they are in a hurry to be pollinated between the early morning when they open and the evening when they are over. All kinds of insects come to visit them and to take the pollen, which is produced in great quantities by the large number of stamens: beetles, flies and bees are the chief visitors and will often be found in the flowers at the same time, together with bugs and any other creatures that can crawl around the open blossoms. Bees have their own special way of handling poppies and similar bowl-shaped flowers with large numbers of stamens; after landing they turn on to their sides and in what looks like a state of high excitement they scramble round the blossom, pulling the stamens between their legs to get the pollen from the anthers and pack it into their pollen baskets. With such short-lived flowers as the poppy the male and female parts have to be ready simultaneously as soon as the blossoms open. With other similar pollen flowers there is often a chance for the two sexes to be ready at different times, as in the case of the Wood Anemone *(Anemone nemorosa)*, another promiscuous species with a varied insect clientele. When the flower opens the stamens are at first bent over the stigmas, and do not move outwards to uncover them until after the pollen has been shed; the stigmas are then able and ready to be pollinated.

Simple and somewhat rudimentary flowers of this kind, with a large and indeterminate number of stamens, usually produce great quantities of pollen, so they can afford to have a considerable proportion taken for food, and indeed simply wasted by primitive beetles and their kind. There tend not only to be messy eaters but to lie about in the blossoms instead of keeping on the move, not being driven by the compulsion to work that activates the social bees and their like. Even after losing a large proportion of their pollen in this way, such species usually have enough grains left over to fertilize them satisfactorily. More advanced flowers, however, with a definite number of stamens and a limited quantity of pollen, may adopt devices to cut down the amount of waste. Some species have two kinds of anther, one conspicuous and offering insects pollen to take away as food, and the other unnoticed and producing the fertilizing pollen which powders the visitor before it goes on to the next flower. One such is the very beautiful pink flowered Crape Myrtle, *Lagerstroemia indica*, which has striking yellow feeding stamens, eagerly visited by insects for the pollen, and drab fertilizing stamens hidden by the extraordinary frilled petals. Another example is the well-known purple flowered shrub *Tibouchina*, which goes to the other extreme by making the fertilizing stamens large and purple, so that visitors cannot distinguish them from the petals, and the feeding stamens bright yellow so that they stand out in color from the rest of the flower.

In some cases the difference between feeding and fertilizing stamens extends to the pollen itself, and many amount to a fraud on the visitor. In the species *Tripogandra grandiflora* the pollen offered by the feeding anthers has become degenerate and infertile; in many members of the Spiderwort family *(Commelinaceae)* some special stamens bear highly colored anthers which produce no pollen at all, so that a visitor is cheated completely unless it eats the anthers themselves, which contain a milky liquid.

Some plants produce imitation pollen. Species of the orchid genus *Eria* carry a powder closely resembling pollen on the lip of the flower, and another orchid of the genus *Maxillaria* produces similar imitation pollen which is eaten by certain bees, as shown by the finding of some in their intestines. Plants of completely different families also use imitation pollen to attract visitors, such as a species of *Rondeletia*, tropical evergreen shrubs, which produce in and around the throats of their flowers a quantity of material very like pollen grains and rich in protein and starch.

Several examples are reported of the way vibration is used to bring about the emission of pollen from the male organs of flowers. In species of *Cassia*, yellow flowered shrubs of the pea family, there are separate feeding and pollinating stamens. The flowers are visited by a large and powerful bee which "milks" the anthers of the feeding stamens and as it does so rapidly vibrates its wings; as a consequence a cloud of pollen is ejected from the fertilizing stamens and some of it settles on the back of the insect although it is sitting on top of the blossom. In the case of the Cow-wheat, *Melampyrum pratense*, the position is reversed; one kind of visiting bee pushes itself into the two-lipped yellow flower, thrusts itself upside down under the overhanging anthers, vibrates its wings and so causes the pollen to be shaken out of the anthers onto its abdomen, where it is collected. It is not only bees that act as vibrators in this way, but also hawk moths and even hummingbirds, which create a whirlwind with their rapidly beating wings as they hover over the flowers.

After pollen, the original attraction for insects, the second powerful inducement for visitors is nectar. A great deal of argument has gone on about when and how the production of nectar first started. Pollen, or microspores, plays a vital part in the life history of nearly all plants. Nectar, on the other hand, performs no function in the actual process of reproduction, but simply attracts pollinators. It is an extravagant use of material built up by the plant. Some people have suggested that in some mysterious way the plants themselves made the first move by inventing nectar as an act of generosity, to reward visitors for their efforts and to persuade them to continue their work of pollination by offering further rewards. Francé, in *The Love Life of Plants*, even went so far as to write: "Who bred the flowers up to secreting honey? Nobody. This is a free and independent deed. . . . First the plant had to make advances, then the insects observed it . . . it was an act of mutual assistance instead of the fight for existence; a reward to the insects for their intelligence and kindness."

Most modern botanists would reject any such explanation, which credits plants not only with the intelligence to think up the idea of giving rewards, like a parent giving candy to a child for helping in the home, but also with feelings of affection and gratitude. A less sentimental explanation is now generally accepted. There is plenty of evidence that, as with pollen and the other types of spores which preceded it, sugary substances very like nectar had been produced by plants of quite primitive types for millions of years before the higher plants arose and brought the need for insect pollination. To this day, some ferns produce a type of nectar which is eagerly sought by insects, including bees, in the early spring before there are many flowers out; yet ferns have no flowers or pollen, so the nectar they produce can have no possible purpose in attracting insects to bring about pollination.

It seems more likely that the production of syrupy solutions started quite independently of flowers, simply as the exudation of surplus sugar. Young shoots of members of the cactus family can often be seen covered with glistening drops of syrup; species of the Prickly Pear cactus, *Opuntia*, will even produce nectar from the tips of their sharp spines. Many other plants also exude nectar from leaves, stems and other places besides the flowers; and such sweet liquid – called *extranuptial*, meaning "outside marriage," because it does not lead pollinators to the sex organs – is taken by all kinds of insects which happen to find it. Ants in particular are attracted by it and often swarm greedily all over it. This can have two advantages for the plant. Ants are notorious thieves, and can rob a flower of nectar without pollinating it, because they are small enough to raid it without touching the sexual organs. So anything that lures them away from the blossoms is a good thing for the plant, and the extranuptial nectar does this very well, particularly as ants seem to prefer it to the floral nectar.

The other advantage is that the extreme fierceness of ants can be put to good use by plants to provide them with highly effective protection. Ants are always ready to bite and sting an enemy, and a plentiful supply of nectar will attract an army of them to frighten off other insects. This is particularly useful for certain plants, such as the beautiful blue flowered climber *Thunbergia grandiflora* from northern India, which is pollinated by large bees. The nectar produced by the flower is found at the back of a long corolla, and to get at it the bee has to force its way into the blossom; in so doing it comes in contact with the sexual organs and brings about pollination. The entrance to the flower is rather narrow and the corolla is thick and strong, so the bee's performance requires a considerable amount of effort. It is much easier for the bee to go round to the back of the flower, bite a hole in its base and get at the nectar that way; but that is no good to the plant, since the bee will have stolen the nectar without doing its job of pollination. The problem has been overcome by means of a number of nectaries behind the flower, covering the calyx and the stalk. These secrete large amounts of extranuptial nectar, which attracts a swarming mass of ants. Any bee that tries to get into the flower by the back way is very soon put to flight by the ferocious gang of ants; if it wants the nectar inside the flower it has to behave properly and use the front entrance. There are many other examples of these "ant guards," particularly in the tropics, where large and destructive insects abound.

Nectar, then, is quite commonly produced extranuptially, and certainly existed in some form long before flowering plants came into existence. But it is within the flower that the production and provision of nectar has reached its highest development. However, it is within the flower that the production and provision of nectar has reached its highest development by adaptation to the interplay between the sexual needs of the plants and the tastes and desires of their visitors. Nowadays nectar is the most important attraction to the largest number of creatures that are able to bring about pollination. In many cases it is offered quite freely to all comers in the insect world, the nectaries being openly displayed and easily accessible in flowers of a primitive, bowl-shaped type, such as the Meadow Buttercup *(Ranunculus acris)*. In simple, open flowers of this kind, the nectar is available to any insect, even the most rudimentary such as bugs and thrips, that can lick, suck or otherwise imbibe it. Development from those simple, promiscuous flowers has come from adapting to the personality, physique, strength, intelligence and habits of more specialized pollinators, including not only insects but other creatures such as birds and bats. Usually this development has been towards more elaborately structured and clothed flowers, concealing their attractions from the common run of casual visitors and making them available only to those prepared to enter into a deeper relationship. The word deeper has a literal significance, because in general the trend of development has been from the open, flat blossom to the closed one with depth to it. Relationships between flowers and their visitors could be said to have developed during the course of evolution from the shallow to the deep.

Before we leave the subject of nectar as a means of attraction, let us look at the way in which it is produced. Floral nectar is normally secreted by a *nectary*, which is a specialized structure made up of closely packed cells surfaced by a more or less porous skin through which the nectar is exuded by a kind of squeezing action as the pressure of fluid builds up in the cells beneath. As would be expected, the surface of the nectary is generally shiny and glistening; the color is usually yellow tinged with green. In the Spurge family the nectaries are very easily seen because there are no petals on the flowers to obscure them; a number of small male flowers, reduced to stamens only, is grouped around a single female flower within a cup-shaped structure called a *cyathium*. The production of nectar is so copious that in one species, known generally as the Poinsettia (*Euphorbia pulcherrima)*, birds come to help themselves to it, attracted also by the bright red leafy bracts, looking like vivid petals at a distance.

Other attractions, both real and deceitful, are offered by flowers besides pollen and nectar, but we will leave them to another chapter. Before finishing this chapter, however, we must deal briefly with two methods of pollination which, because they do not involve the exploitation of living creatures, need no attractions at all.

One we have already touched on is pollination by wind, technically known as *anemophily* ("wind love"). In this case, evolutionary adaptation has led plants in exactly the opposite direction to that of those which rely on live carriers to bring about their mating. Instead of putting on more elaborate or conspicuous clothing and concealing their organs more thoroughly, wind-pollinated flowers have tended to strip themselves naked and expose themselves as much as possible. The male sexual organs have become long and thin, and hang down so that the least puff of wind or even current of air will make them swing and dance, emitting pollen in clouds. The female organs are large and hairy, the styles and stigmas often being elaborately plumed so as to catch every possible grain of pollen before it falls to the ground and is lost. The nature of the pollen is usually quite different from that of plants that rely on living things for pollination: because it has no need to cling to the coats of visitors, it is not sticky and patterned but dry and smooth, so that the grains do not stay together but are dispersed as widely as possible.

Wind pollination can, as we have noted, take vast quantities of pollen; a single hazel catkin can produce four million grains and a birch catkin five and a half million. Obviously the waste cannot be excessive or such a large proportion of the plant kingdom, including the grasses and many of the trees, would not continue to be wind pollinated.

The other form of pollination which is independent of living things is pollination by means of water, technically known as *hydrophily* ("love in the water"). It is rather rare in the higher plants, which are very far removed from the algae and their primitive kind with which life began in the sea.

The Tapegrass, *Vallisneria spiralis*, has ribbonlike leaves and leads a submerged life in rather shallow water. The two sexes are on different plants. The female flowers start as buds borne on very slender stems coming from the base of the plant. These gradually elongate until the buds reach the surface of the water; during their journey upwards these buds are protected from the water by a watertight covering. When they reach the surface the buds open out to expose three large stigmas, thickly covered with hairs which repel water. The open female flowers then float on the surface, waiting for the males to come along. The male flowers are very much smaller and several together are borne in a sort of tubular envelope near the base of the plant. When they are sexually mature, these tiny male flowers break away from the plant and float up to the surface, each tightly shut to protect the contents from the water. When they reach the surface they open, the perianth segments bending right back to support the flowers as they float, and exposing the two (or occasionally three) stamens. The anthers open, revealing a mass of ripe pollen, and the male flowers float around on the surface till they chance to come near a female flower. Small depressions in the surface film around the female flower cause the male flowers to be pulled towards it, and the stamens touch the spreading lobes of the stigma, which transfers the pollen to its hairy surface. Once pollination has been achieved, the female flower is pulled down below the surface by means of a spiral coiling of the stem, which causes it to resemble a spring.

The development after fertilization therefore takes place under the water, but it will be noticed that the actual pollination, that is the transfer of pollen to stigma, takes place in the air. The same is true of the majority of aquatic plants. Even in those cases such as the Canadian Waterweed, *Elodea canadensis*, a relative of the Tapegrass, in which the anthers of the male flowers explode, scattering the pollen over the surface of the water, it is found that the outer coat of the grains is so patterned that it keeps them from getting wet and allows them to be buoyed up by tiny air-bubbles as they float around till they meet a female flower.

In a very few cases, pollination does actually seem to take place under the water, as with Hornwort *(Ceratophyllum)* and Eelgrass *(Zostera)*, in which the pollen, instead of being in rounded grains, is in the form of tiny threads, which drift on the tide and instantly wrap themselves around the narrow stigmas as soon as they come into contact with them. In this way the male and female avoid being torn apart before their mating is completed.

VII

Exhibitionism

To human eyes perhaps the most blatant example of what looks like sexual exhibitionism in the plant world is the fungus *Phallus impudicus*, which means "shameless penis." As can be seen from the picture at the beginning of this book, the name is an apt one. The Stinkhorn, as the fungus is commonly known, is remarkably similar in shape and size to a rather well-endowed human penis in a state of erection. It appears in the late summer as a bulge in the ground and quickly swells to the size of an egg; the tip then breaks through the skin and within a few hours it has elongated to full size.

Just as striking as the appearance of *Phallus impudicus* is its smell. In her book *Rambles in Search of Flowerless Plants*, published in 1865, Margaret Anne Plues tells how a friend walking with her through the Swaledale woods in Yorkshire thought that some rabbits or larger animals must have died, because the stench of rotting corpses was so disgusting. "As we approached, the odor he complained of became painfully perceptible. Another turn in the path, and I beheld a group of *Phallus*, tall and stately, like a group of marble obelisks. I hastened to assure my friend that the 'deceased animals' were no others than living Stinkhorns."

The way the fungus attracts the visitors it needs is not by its phallic appearance but by its stench, which may be sickening to humans but is delicious to carrion flies that live on putrefying flesh. The source of the stink is a dark green slimy jelly, which covers the pointed knob on the end of the shaft and which is greedily eaten by the flies that swarm over it. In that smelly jelly millions of spores are embedded, and these are swallowed by the flies along with the rest of their slimy feast. Being highly resistant to digestive processes, the spores pass through the flies unharmed and are expelled with their excreta. A single speck excreted by an ordinary carrion fly can, it has been estimated, contain twenty million spores, which rely on this method of distribution to produce further generations of Stinkhorns.

There is another species, called *Phallus hadriani* ("Hadrian's penis"), which is rather thinner, usually somewhat bent, has pinkish flesh, gives off a sickly sweetish smell, and in addition to flies attracts a good many slugs, which respond eagerly to the odor of decay. A smaller member of the same family is called *Mutinus caninus*, the first word coming from Mutinus, another name for the ancient god of male potency, Priapus (still commemorated in the old medical term priapism, used to describe a condition marked by persistent erection). The second part of the name, *caninus*, meaning doglike, is a perfect word for the fungus, which looks exactly like a dog's penis, from its somewhat flushed base to its shiny, bloodshot-looking head. Commonly called the Dog's Stinkhorn (and sometimes ruder names), this fungus succeeds in smelling even more repulsive than its two relatives just described. It combines the putrid stench of the first and the sickly sweetness of the second with an added tang of rotting excrement. Though not quite as strong as that of *Phallus*, the stench that comes from this combination of smells is indescribably foul and draws dung-flies and their like from a long distance, to gorge themselves on the slime and so distribute the spores.

But what may look to us like sexual exhibitionism on the part of these fungi is really nothing of the kind. How they appear to us is of no importance to them, since they have nothing whatever to gain from the impression they make on human beings. And as for the flies, they can hardly be expected to respond to the phallic appearance; it seems most unlikely that an imitation penis, whether of a man or of a dog, could excite them. We must therefore assume that since the impersonation brings no benefit to the fungi, the strong resemblance to a penis is purely a matter of chance.

Another member of the same family, the *Phallaceae*, does put on a striking and colorful visual show which has defied many attempts by scientists to explain it. Its name is *Clathrus ruber*, and it is commonly called the Latticed Stinkhorn. Like *Phallus*, it makes its appearance as a swelling "egg" with a dirty white skin. As this expands it begins to show signs of being stretched over a ribbed network. When the skin can stretch no more it cracks, and with remarkable speed there bursts through it a strangely beautiful structure in the form of a hollow sphere, perforated to make a grating through which flies can wander in and out. The most surprising thing about this latticed ball is that it is colored bright coral red, almost as if it had been painted. So extraordinarily beautiful is it that collectors have gathered it and taken it home; but, to quote from M. C. Cooke's *British Fungi*, "It is of so putrescent a nature that its odor materially detracts from its beauty; it is recorded that a botanist who gathered one was compelled by the stench to rise during the night and cast the offender out at the window."

Just as the Latticed Stinkhorn outshines all the other members of its family in visual beauty, so it appears to outdo them all in the sheer sickening nastiness of its smell, which is given off by olive brown mucus thickly spread over the inside of the spherical grating. The violent contrast between the appalling stink of *Clathrus* and its coral-like appearance clearly fascinated Francé, who wondered whether the color was put on like lipstick to attract callers. In *The Love Life of Plants* he wrote: "A little fungus, as red as fire, stands up before us and boldly asserts its right to be called a flower, for she has a little flirtation with the fly." Another of the *Phallus* family whose extraordinary visual beauty struck Francé was the fantastic *Dictyophora*, which like *Phallus* closely resembles a human penis, but with a remarkable addition: around the shaft, just below the head, is what appears to be a rather elaborate lace collar, or perhaps a surplice, made from a network of membrane called the *indusium.* In spite of its resemblance to an upstanding male organ decked in a frilly accessory, it is referred to by Francé as "The Lady with the White Veil, which shoots up in the late afternoon within two hours, unfolds all at once the most delicate lace veil, receives her visitors during one night, and ends at dawn a little heap of rubbish."

Many other of these foul-smelling fungi are, he reminds us, visited at night, when numbers of creatures that feed on corpses are active; some of the fungi even phosphoresce in the darkness "like little ghosts. . . . They flirt at night, and in order to be found by their admirers they hang out a lamp. That is to say, they produce light."

Few botanists nowadays would believe that the visual display of these stinkhorns, strangely beautiful though it is, plays any part in attracting visitors. Just as flies are unlikely to be interested in a mock penis, so they cannot be excited by the coral color of *Clathrus*, since they are unable to see red; they do not seem to go for the lace collar of *Dictyophora*, since the slime they enjoy is confined to the head; and the dim nocturnal light given off by some species does not appear to bring them any more visitors than the unlit ones; in any case the stink is attraction enough, without the need for any of these visual tricks.

The exhibitionism of the stinkhorns is in fact a matter of smell rather than sight. And it is all based on deceit, like so many of the lures which plants use to attract callers. The smells promise the flies a meal of putrid flesh or rotting excrement, but the promise is a lie; what the flies get is not flesh or excrement but slime. However, the attraction, though deceitful, is not entirely fraudulent. At least the insects do get a meal, and one they thoroughly enjoy, because they greedily eat up every particle of mucus, leaving the fungus bare and quite without smell, which is entirely confined to the slime.

In many other cases of olfactory exhibitionism, where deceitful smells are used as enticements, the whole thing is a total fraud, and the deluded visitor gets nothing, not even a substitute. We will look at some examples later, in the section on bondage and sadism. Meanwhile it is important to remember the crucial part often played by smell in the insect's final decision whether to visit a flower or not; experiments have proved that however strongly the caller may have been drawn by visual attraction, when it gets within touching distance what may well make its mind up whether to stop or fly on is whether the flower smells right or not.

Before moving on to other forms of exhibitionism, let us glance at two particularly smooth confidence tricks in which misleading scent is used as bait. The first sucker – literally and figuratively – is a small blood-sucking midge. It was long wondered why the blossoms of *Arum conophalloides* (meaning "like a conical penis") were visited only by female midges. What was the exclusively female attraction? Then it was realized that only the females suck blood, which they need to complete their sexual cycle, and that the blossoms were giving off just the same smell as the skin of the animals on which the midges usually feed. The suckers got no blood from the visit, but the plant got everything it wanted in the way of sexual satisfaction.

The other plant that uses deceitful smells to defraud visitors is *Stapelia*, of which some ninety species are found in tropical and southern Africa. They are succulents, very like cacti to look at, and they live in hot places where flies abound. The flowers, which appear either singly or in clusters from the thick stems, are very strange in appearance; in most species they look rather like starfish, with fleshy segments which are purplish brown in color and have the appearance, texture and smell of bad meat. Experiments have shown that though carrion flies prefer yellow flowers in the ordinary course of events, they immediately change their preference to purple and brown when there is a stink of putrid flesh about; the combination of the appearance and the smell of bad meat is quite irresistible.

To make the deception even more convincing, the flowers of most species of *Stapelia* are flecked and covered with hairs, so that they look moldy; they are also edged with a fringe that is moved by the slightest air current, giving the appearance of a teeming mass of small flies already at the meat. It is not simply greed, however, to which the deception appeals, but also to the maternal instinct. When a fertilized female carrion fly is ready to lay her eggs, she does so on a piece of rotten meat, so that as soon as they hatch out the maggots will have an appetizingly putrid meal waiting for them. So convincingly does the *Stapelia* flower resemble rotting flesh in color, texture and smell that female flies lay their eggs on it. In doing so they get the pollinia, which are rather like those of orchids, fixed to them by a sort of clip, and these remain till they are wrenched off by the stigma of the next flower. The fraud in this case is perpetrated not only on the female fly but more particularly on the next generation; if the eggs hatch out, the maggots will find that the flower has no food to offer them at all, and they will die of starvation.

There are a great many more examples in the plant world of deceitful smells, that is smells which imitate other odors. In most cases, though, the scent of flowers does not pretend to be anything other than itself; it is what Professors Faegri and van der Pijl call "absolute odor," in other words something that has no counterpart outside the sphere of blossoms. Its function is to attract the pollinator to the flower; it offers no satisfaction in itself, but leads to a source of satisfaction, usually in the form of nectar or pollen or both.

Recently, however, a most remarkable discovery has been made. The perfumes of flowers can be sought by some creatures for their own sake: not because those perfumes promise, genuinely or fraudulently, other satisfactions such as food or drink but because they offer in themselves the satisfaction of a different kind of craving. It was observed that certain male bees actually have special organs on their hind legs in which they collect the scent from flowers in the form of drops of fragrant oil. The males go from flower to flower using their front legs to scoop up the drops of perfume, which they then transfer by way of their middle legs to the scent containers in their hind legs, filled with hairs like cotton wool which soak up the oil. During the process of collection a male often becomes quite drugged. After he has filled up his perfume containers, he then stations himself in a solitary position, where he gives off the scent he has just collected. It is thought that this peculiar performance is based upon the sex urge; the male is behaving exactly like a man who has used the latest after-shave lotion, which the advertisements claim is irresistible to women, and is now waiting for the magic to work and females to come flocking to him.

This collecting and use of "male toiletries" has been observed in many species of orchid, which produce the perfumed oil in considerable quantities; it has also been seen in flowers of quite different and unrelated families. It may be quite a common practice, though only recently noticed; it may even be a growing fashion. Whatever new discoveries may be made about it, it calls into question many previous observations of male bees taking drugs from flowers apparently for their narcotic effect. Now it is thought that some of this drug-taking may be indulged in not merely for its own sake but as a means of attracting females and getting them in the mood.

Exhibitionism by visual display is widespread throughout the plant world; it is indeed responsible for all the beauty of flowers, which as we have seen has an entirely sexual purpose. Sometimes in the interests of the most effective display other parts of the flower than the petals may take on the role of attracting visitors. One example is the Clematis, in which it is the sepals, normally small and green, that have become large and brightly colored. True petals are entirely lacking in this flower, though among the numerous stamens there are usually some reduced organs called *petaloid staminodes*. The enlarged sepals, or *perianth segments*, mimic petals exactly and usually carry guidelines, often embossed as well as colored, to steer insects towards the middle of the flower, where the sexual organs are.

Generally the sexual organs themselves play little or no part in visual display. They are the vital parts on which the next generation depends, so they need to be protected from harm; hence in a large number of cases they remain modestly unobtrusive and leave the publicity to the accessory parts, whose job it is to put on a show. Sometimes the male organs are fully exposed, especially when there are large numbers of them, so that several stamens can be damaged or even destroyed without materially diminishing the amount of pollen produced. There are even cases where the male organs themselves are decorative and provide the main visual attraction. One example is the Common Meadow-rue *(Thalictrum flavum)*, a member of the buttercup family which carries massed clusters of small flowers;

the tiny whitish perianth segments are inconspicuous, but the long, erect stamens make the clusters of flowers look golden yellow and attract multitudes of hoverflies, honeybees and bumblebees, which come for the pollen, since there is no nectar. The stamens are so long and the pollen so copiously produced that pollination by wind often takes place; since the pollen is smooth and dry, so that it does not stick particularly well to insect visitors, and since other species of *Thalictrum* are known to be anemophilous, it may be that the Common Meadow-rue is in the process of evolving into a wholly wind-pollinated species like its relatives, and that the long stamens, which at present give rather unstable support to the heavier visitors such as bumblebees, will become longer still and more suitable for wind than for insects.

Perhaps the most spectacular examples of flowers that not only expose their male organs but make them colorful and decorative are to be found among those Australian members of the myrtle family generally known as Bottlebrushes. Their botanical name is *Callistemon*, which comes from the Greek for "beautiful stamens," and never was a plant more aptly named. Anyone who has seen a clump in full bloom in the Australian bush will never forget the sight. As the common name suggests, the densely packed spike of flowers exactly resembles in form one of those cylindrical brushes with stiff bristles used for cleaning bottles. One of the most common species, and perhaps the most beautiful, is *Callistemon citrinus*: its bristles are in reality the stamens, which are long, erect and colored the most brilliant crimson. Its visitors are birds, which have much the same eyesight as humans; unlike bees and most other insects, they are blind to ultraviolet light but are strongly attracted to red, the brighter the better. Having little or no sense of smell, they rely heavily on the vivid color to guide them to the nectar which the flowers produce copiously and which the birds consume greedily to replace the energy they burn up during flight.

The female sex organs are usually even less likely to make an exhibition of themselves than their male counterparts. Their role is to produce offspring, not to show themselves off. However, there are some cases where the female sex organs are not only openly displayed but also enlarged, colored and even ornamented with frills, so that they take over much of the advertising in addition to their normal sexual function. Perhaps the most striking example is the Iris, which has three large and colorful lower perianth segments (sepals) often bearded and called the "falls," and three other (petals) standing erect above them and called the "standards." Up through the middle of the flower comes the highly decorative female organ, the "crest," divided into three spreading style-branches, each with a double frill at the end hiding the stigma underneath. Each branch of the crest comes above one of the falls, which carries a pattern of veins forming a nectar-guide and leading inwards towards a supply of nectar at the angle where the crest and the falls meet. Directed by the guides, a bee will crawl under the crest and along one of the falls to get the nectar. In doing so the insect will be powdered over with pollen from a hidden stamen just above its back. When the insect repeats the performance with another flower the pollen will be combed off its back by the stigma hidden beneath the frill at the end of the crest and so will have earned its feast of nectar by bringing about pollination.

At the opposite extreme to flowers that expose their sexual organs, and even decorate them, to attract attention, are those that have given up sex entirely in order to devote themselves to luring visitors to other fully sexed flowers which lack the personality to lure visitors themselves. In the daisy family there are many examples of such procurers – sexless themselves, but good at making sexual arrangements for others. Several species have big, bright and attractive, but sterile, ray florets around the edge whose sole job is to attract insects. These wander over the central part of the daisy and pollinate the dull and unobtrusive, but sexually potent, disk florets.

The same technique is found in other families too. A species of Grape Hyacinth, *Muscari comosum*, of the lily family, has very attractive violet colored flowers at the top of the raceme and dull brownish ones below. Insects are drawn to visit the top flowers, which are not only sterile but contain no nectar; however, in most cases the visitor will work its way downwards and soon come to the dull lower flowers, which contain both sexual organs to gain from the insect's visit and nectar to repay it.

A more spectacular example of a plant that employs sexless but attractive flowers to drum up patrons for sexual but dull ones is the Guelder-rose *(Viburnum opulus)*, which bears heads of pallid, insignificant, fertile flowers surrounded by a border of sterile but much larger flowers, sparkling white and very attractive, both to insects and to humans.

Perhaps the best known examples of the use of sexless flowers to attract attention is the Hydrangea. The lacecap varieties, commonly grown in gardens, have large flat heads of blossoms, made up of showy, pink or blue sterile flowers surrounding a central group of small fertile ones. Some of the cultivated varieties, raised from cuttings and sold by the million for house and garden decoration, have lost their fertile flowers completely; their large, round corymbs, looking like great plates of blossom, consist entirely of showy sterile flowers with no sexual organs. What started as a device to attract visitors for sexual purposes has ended in complete sexlessness. As the final irony, these sterile flower heads no longer attract insects, because they have nothing to offer but show.

VIII

Bondage and Sadism

Like those people who cannot perform sexually unless they are tied up and beaten, there are other living things that seem to need some degree of sado-masochistic stimulus before they are aroused sufficiently to be capable of sexual intercourse.

There are several cases in the animal world where considerable violence appears to be an absolutely essential preliminary to mating. Teeth and claws and sheer brute strength are often used before the more passive partner – usually, but not always, the female – finally submits. Some people argue that all sexual behavior contains an element of aggression, involving as it does the hunter and the hunted; even the use of persuasion by some animal species, including man, to gain the desired partner's willing consent is said by some to be only a subtler form of coercion, with gifts or smooth talk or attractive display taking the place of physical force. There are many examples of these more seductive methods, but at present let us confine ourselves to violent sex.

One of the most striking examples of sado-masochism to be found in the world of nature is the courtship behavior of the snail – if courtship is not too tender a word for such a rough affair. We have already noted that the snail is bisexual. As with hermaphrodite plants, this arrangement has the great advantage that every encounter between any two members of the species can lead to mating, whereas if there were different males and females half the encounters would be homosexual and lead nowhere. Copulation is slow and long drawn out, each partner simultaneously acting as male and female. It may sound like the last word in thrilling sexual experiences, but before the slow moving, cold blooded creatures can work up any interest in sex they have to be stimulated, and this takes some doing. Nothing less than violence and physical pain can arouse the snails sufficiently for them to be able to perform. In the male reproductive organs, near the penis, is a special sac in which is produced a sharp, flinty spike called the *spiculum amoris* ("love's dart"), which is fired with considerable force into the partner. The dart pierces the partner's tender skin and stings it into response, often by firing its own dart into the first one's body. Only after the shock of this painful initiation into sex can lovemaking proceed. The aroused couple begin to feel each other with their "horns," and a long session of mutual stimulation follows, ending in a remarkable bisexual performace which is too complicated to be described here and which would be illegal in many human societies.

There is nothing quite like that between plants, but there are many instances of bondage and cruelty in the relationship between flowers and their visitors. The nearest approach to masochism by a plant is probably to be found in the case of Dyer's Greenweed *(Genista tinctoria)*, a member of the pea family with yellow flowers. It is visited mainly by rather large bees, which need not only size but strength to force their way into the blossoms. The parts of the flower are so arranged that when it opens they are under considerable tension. As soon as the first heavy visitor thrusts its way in, the weight rips the flower open with explosive force, leaving only the top petal erect, with the sexual organs pushed against it, and the other four petals hanging limply down. During the explosion the visitor is peppered with pollen. The violent deflowering leaves the blossom completely wrecked, unable to repeat the performance and with its looks gone for ever. What is not clear, and perhaps never will be, is whether the visitor sets off the explosive mechanism in an attempt to find the nectar that many similar flowers contain; if so, the insect has been fooled, because Dyer's Greenweed has no nectar at all.

Another member of the same family has managed to avoid a similar fate and to keep its shape and remain attractive after being deflowered. It is called *Clitoria*, because the flower looks very like a woman's sexual organs with a rather well developed clitoris, as can be seen from the picture at the beginning of this book. The way the *Clitoria* flower has saved itself from damage by big, rough insects is to turn itself upside down, so that the broad petal normally at the top in this type of blossom is now below, making it a strong landing place for even the heaviest bee. The "clitoris" formed by the other petals is not required to bear the visitor's weight and so is not hurt or deformed when the insect thrusts itself forward, probing with its tongue to reach the nectar inside. It is a strenuous performance, needing the bee to brace its legs against the landing platform and exert considerable force, but by being upside down the flower is able to take the strain and to yield to the visitor just enough for it to gain entry but not so much as to be ruined in the process. The grip of the floral parts delays the withdrawal of the bee so that it becomes thoroughly powdered over with pollen before getting free and passing on to the next flower.

This enforced delay of the visitor's withdrawal from the *Clitoria* flower is a very mild example of the many techniques of entrapment, bondage and sadism that blossoms practice on their victims. The Butterworts *(Pinguicula)* go much further and often murder their pollinators. These plants grow in marshy or boggy places, the sort of places that abound in small flies. The species *Pinguicula alpina* carries white flowers on stalks above a rosette of greasy looking leaves; the entrance of each flower is ornamented with two little bumps covered with bright yellow hairs, which attract flies. These crawl right into the flower and stick their heads into a hollow spur at the back. This looks as if it might contain nectar, but it does not; instead it is lined at its lower end with small hairs topped with round, juicy heads on which the visitors feed. When a fly has had enough, it tries to pull itself out again backwards, but it finds withdrawal a great deal more difficult than entry. The reason is that at the entrance of the spur there is a thick clump of inward-pointing hairs. These gave the fly no difficulty when it was pushing its way in, because they were lying in the same direction as it wanted to go; in fact they made a nice soft resting place for it while it was eating. But when the fly tries to pull itself out, the hairs tug at it and prevent its withdrawal; and the harder it struggles to get free the more tightly the hairs grip it and block its exit. The only way the fly can get out is by slow and careful withdrawal, which means that it has to have both the intelligence and the strength to push its body upwards so as to get free from the hairs which are grasping it by the belly. In doing this it touches the flower's sexual organs with its back, and since the blossoms are protogynous – that is, they start female and change to the male state later – the fly brings about cross-pollination by carrying the pollen from the older flowers to the female organs of the younger ones.

What, though, if a fly is too weak, too clumsy or too stupid to succeed in pulling itself out of the flower? The answer is that it remains stuck there, its struggles getting feebler and feebler, till it dies of starvation. In fact the Butterworts seize and kill insects in two different ways. On the one hand, as we have seen, they grip flies with their flowers, and if the flies are strong and clever enough to get free they are used as mating agents, otherwise they suffer a lingering death. On the other hand, the leaves are covered with a clammy film, secreted by special glands, which traps all kinds of small insects; their struggles to get away cause still more sticky juice to be secreted, together with an acid fluid which quickly digests them. Many other bog plants also act as "living flypapers," trapping insects by means of sticky secretions and then digesting them, so making up by a meaty diet for the lack of nourishment in the soil of such places.

A particularly cruel type of trap to catch insects and force them to act as pollination agents has been evolved by the Milkweed family *(Asclepiadaceae)*. It is known as a pinch-trap, and although the details of its construction vary from species to species its method of operation is simple, effective and unscrupulous. Like orchids, members of the Milkweed family produce their pollen in waxy masses, called *pollinia*, instead of in a fine powder. This means that the flowers cannot simply dust visitors over with pollen but have in some way to fix the pollinia to them securely enough to stay in position till they reach the next flower. Unlike orchids, which for the most part have solved the problem by sticking their pollinia to insects by quick-setting glue, the milkweeds use clips. These are made of a hard plate of horny material with a slit underneath, narrowing upwards. Each clip is fixed by two strong bands to a pair of pollinia, contained in pockets around the column. The way the clips hook themselves on to insect visitors, which are then held tightly down by the strong bands, has caused at least two species in the family to be called "cruel plants."

One is *Cynanchum acuminatifolium*, also called the Mosquito Plant, which has clusters of small white flowers; these are visited by flies and other small creatures, lured by the cups of nectar around the top of the blossom. When an insect after drinking the nectar tries to fly off, it is likely to get its leg caught in one of the clips, and the more it struggles to get away the tighter the clip fastens itself on by the narrowing slit. If the insect is strong and agile enough it will tear the pair of pollinia out of their pockets; by a remarkably quick twisting movement the two bands bring the pollinia together, so that when the insect visits the next flower they will slide easily into a slot specially designed to receive them and wedge themselves against the stigmatic surface. The only way the insect can get away is by wrenching the bands till they break away, leaving the pollinia behind. A great deal of strength and energy is needed for the visitor not only to lift itself off by its wing power but to snap the bands holding it back. Some visitors, especially the small flies, cannot make it and remain helplessly fluttering, till finally they flutter no more.

Most of the visitors to the smaller flowered members of the milkweed family are flies, whose entrapment, efforts to escape, suffering and sometimes death tend to go unnoticed. Some of the species of milkweed with more spectacular flowers attract bigger and stronger visitors, such as the larger bees, which come to suck nectar from the prominent cups of the beautiful North American Silkweed *(Asclepias syriaca)*, with its clusters of rosy purple flowers. A close relative, *Asclepias tuberosa*, with heads of bright orange flowers, attracts crowds of butterflies, gently fanning their colorful wings hour after hour in the sunshine, drinking the copious supply of nectar in the hoods which crown the blossoms.

Very rarely do these larger and stronger insects get caught by any species of *Asclepias* so that they cannot manage to wrench themselves away in the end, even if it takes a struggle to do so. Another member of the family, *Araujia sericofera*, which comes from the warmer parts of South America, does however claim a considerable number of victims of its lethally efficient pinch-trap. It is another member of the family to be called the Cruel Plant, and it richly deserves the title. It has clusters of white, fragrant flowers, very attractive to night-flying moths, which do not land on the flowers but hover by them, whirring their wings and inserting their long tongues into the cups crowning the flowers so as to suck up the nectar. The result is that the moth is caught not by the leg but by the tongue, on which the straps fixed to the pollen masses are clipped in a tenacious grip. If the moth is strong it may pull itself away from the flower, with the straps and the pollinia still firmly fixed to its tongue. If it is weak it may flutter there all night, till the sun shines on the flowers coaxing them to open a fraction more, thus allowing the moth to escape, still with the pollen clipped to its tongue. Often, however, the moth is too feeble or too tired to make the effort to get away; its wing beats gradually become feeble and finally it dies, hanging by its tongue from the Cruel Flower.

Another genus in the same family, called *Ceropegia*, adds to the pinch-trap another device known as a pitfall trap. The tube of the flower is lengthened and the lobes are fixed together at the top so that the whole thing looks like a rather elegant lantern. From the corolla lobes comes a scent, rather faint to us but powerfully attractive to certain flies and similar small insects; in some species the top of the flower is crowned with shimmering hairs, which by looking like swarming insects add to the attraction. The combination of smell and sight lures flies to enter the flower, the tube of which, just below the lantern, is lined with tiny smooth hairs, pointing downwards and covered with slippery oil. The flies lose their foothold and slide down the tube into a round chamber below. Unlike the traps we have just looked at, where the strength and ability of the visitors determine whether they escape or not, the trap set by *Ceropegia* gives them no chance at all of getting out by their own exertions; try as they will to climb the walls of their prison, the insects slip down the oily surface onto the floor of the chamber again. A few days later, in its own good time, the tube of the flower droops down to a horizontal position, the hairs inside shrivel, and such insects as have survived their ordeal are able to escape.

It has been found that several species of these lantern flowers attract only female flies and midges; the scents given off are apparently connected with the egg-laying phase of the female sexual cycle. A fly alights on the scent-producing lantern, crawls inside, attracted by the dark interior, slides down the tube and finds itself in a chamber with dark walls except for some partly translucent "windows" in a ring around the flower's sexual organs, allowing in a little dim light. The thoroughly alarmed insect crawls towards the light, thinking it has found a way out, climbs up the central pillar of the flower and finds to its surprise and pleasure that there are cups of nectar waiting for it. After drinking its fill, the insect withdraws its head and promptly gets the pollinia clipped to it; if it already carries pollinia from a previous encounter, these are caught in a groove and wrenched off, getting pressed against the stigma in the process. So the fly has been trapped by the flower in a whole series of stimuli which prey upon its instincts: first the smell and then the dark passage, both exploiting its egg-laying urge; then the light, promising escape; then the nectar, appealing to hunger. After performing its unwitting task of pollination, the insect stays trapped in the flower for sometimes as long as four or five days before the flower droops, the hairs inside shrivel and the fly is able to escape – if it is still alive. As soon as it has recovered from its experience, the stupid fly is attracted by a fresh young flower and gets trapped once more.

There are many other types of trap blossom which are highly efficient at imprisonment and bondage, particularly the various kinds of arum.

These attract flies and beetles that live on dung, dead bodies and other decomposing matter, and to do so they produce the same sort of stench as we have already noted in the case of *Phallus impudicus* and the other Stinkhorn fungi. The most disgusting smell of all, it is claimed, is that of *Amorphophallus titanum* (which means "gigantic deformed penis"), an enormous aroid of the tropics with a large underground tuber from which it sends up a phallic spike, called a *spadix*, which can reach the height of a tall man. From this huge erection there comes a sickening stench of excrement and decaying flesh, with what have been described as the additional odors of rotting fish and burnt sugar. The chief visitors to this super-stinker are large beetles, which, in searching for the non-existent decaying matter promised by the smell, fall into a chamber at the base of the frilled, vase-shaped spathe. The beetles try to escape by climbing up the phallic spadix, but this has an overhanging ridge with such a sharp edge that they fall off it and back into the trap again.

Perhaps the best known of these plants with trapflowers is *Arum maculatum*, commonly called Cuckoo Pint or Lords and Ladies. The most thoroughly studied species, however, is *Arum nigrum*, the Black Arum, from southern Europe. The purplish-black broad blade, the spathe, is lighter at the base, where it forms a hollow chamber. The tall spadix carries female flowers at the bottom, then a ring of bristles, then the male flowers, then more bristles, and above that the rest of the penis-like erection. By an extraordinary mechanism that rapidly burns up stored starch, a good deal of heat is produced, which greatly increases the smell of excrement. Dung-flies, excited by the prospect of a hot meal, crawl inside; the oily walls offer no foothold and the flies fall straight into the chamber below. Once trapped, the insects are unable to escape, both because of the slippery surface and because the bristles bar their way. The female flowers at the base of the spadix carry drops of liquid which the flies lick up, and this brings them to the stigmas. The female flowers remain sexually receptive for a day. The next morning the male anthers open and powder the trapped insects with pollen. The heat dies down, the surface has lost its slipperiness, the bristles have shrivelled and the insects are able to climb up and make their escape. A nearby *Arum*, newly opened and in its female phase, may just be coming into heat and sending out a powerful smell of fresh, steaming dung. The flies will be lured, fall into the trap and transfer their powdering of pollen to the receptive female flowers.

To heighten the smell of decaying matter, some members of the family can produce an amazing amount of heat. *Schizocasia portei*, from the Philippines, has been known to reach 43°C, the temperature of a high fever. The Skunk Cabbage, *Symplocarpus foetidus*, which grows in cold northern regions, can melt snow and create a strong (and disgusting) enough smell to attract insects from a considerable distance.

A great many other types of traps are used by plants. We will look at some more later. Let us end this chapter on a less sadistic note by observing that some of these traps may have originated from flowers that without imprisoning their visitor simply offer them a sheltered place in which they can breed. We have already seen that flies can be induced to lay eggs in flowers that smell of bad meat. Some flowers provide a suitable place for an earlier stage in an insect's sex life, namely the sex act itself. The common Heather, *Calluna vulgaris*, has flowers that are big enough to offer shelter to the tiny thrips which are often found inside them. The male thrips are unable to fly and are much rarer than the females. So the females have to take the initiative. They fly from one flower to another on the lookout for sexual partners. They land on the projecting stigmas, and if there is any pollen on their bodies they fertilize the flowers. The blossoms seem to be regularly used by these tiny creatures as places where they can meet and mate.

Many other flowers offer a quiet, well protected and comfortable spot where visitors can make love to each other. It has been said that it is almost impossible to find some heads of flowers on a sunny day without discovering insect couples in the act. In some inhospitable regions with little shelter of any other kind, flowers are regularly used as courting places. An example is *Lithops*, one of the genera of "living stones," the leaves of which protect themselves from predators by looking almost exactly like the pebbles which litter the ground in their native habitat in southern Africa. From a cleft between the paired succulent leaves appear the starry flowers, and these provide a perfect lovers' meeting place, where insects can find partners, court and copulate.

The realization that flowers can provide at the same time a dating service for insects on the lookout for partners and a private room for them to make each other's sexual acquaintance has come rather late in the study of pollination, and much has still to be discovered in this field. Some extremely interesting observations have been made by Müller in the Alps, where he found many instances of butterflies being attracted by flowers of similar colors to themselves. These apparent examples of self-love were, he thought, really expressions of color preference in the choosing of a mate, and that preference had become transferred to the choice of flowers instead.

Perhaps the process that was eventually to lead to insect mimicry by such flowers as the Bee Orchid all began ages ago when an insect looking for a mate on a blossom and not finding one transferred its attention to the flower instead.

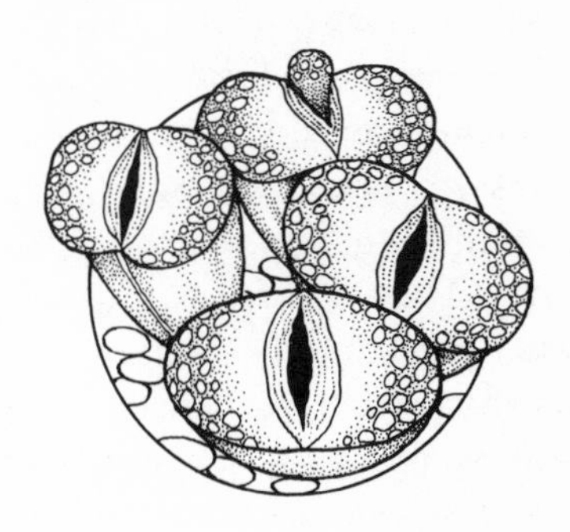

IX

Plant Pregnancy

At exactly the same time as Sir Thomas Millington's discovery of plant sex was being announced in England by Nehemiah Grew, sensational revelations were being made about animal sex just across the North Sea. The great Dutch microscopist Anthony van Leeuwenhoek had taken a drop of fresh seminal fluid and looked at it through his latest improved lenses. The same thing had been tried a few months before by Stephen Hamm, but it was Leeuwenhoek's findings, described and illustrated by him in 1677, that excited the learned people of the day. That tiny drop of semen could be seen under the microscope to be alive with large numbers of swimming, tadpole-like creatures with roundish heads and long tails.

For the first time the male reproductive cells, the sperms, had been seen. It was, however, to be a very long time before it became possible to understand the nature of those sperms or the part they played in fertilization. As we have already noticed several times, orthodoxy dislikes discoveries; if it cannot either brand them as sinful or ignore them, it tries to fit them into its existing systems of belief. And that is exactly what happened in the case of the sperm.

The accepted dogma of female inferiority, justified by many biblical tests and summed up in St. Paul's assertion that man was not created for woman but woman for man, was strongly reinforced by the use of the word seed to mean two completely different things. In addition to meaning the thing from which plants grew it was used throughout the bible for descendants, as in the seed of Abraham, and hence for semen, as in the seed that Onan spilled on the ground. The latter meaning was common at the time; in a book on anatomy published in 1668, only nine years before Leeuwenhoek's discovery, Nicholas Culpepper and Abdiah Cole wrote "Others have attributed to the Kidneys the preparation of Seed, because hot Kidneys cause a propensity to fleshly lust."

Since plants were generally thought to be sexless and females inferior, and since the word seed meant semen as well, it was accepted that the male fluid was exactly like the seed of plants: it contained in miniature the new life complete in every detail, and only needed sowing in a suitable place. The female was that suitable place and her function just like that of the soil, to provide shelter and nourishment, but nothing more. So thoroughly brainwashed were the faithful to accept such beliefs, so incapable were they of questioning holy writ, that some of them managed to find confirmation for those totally false views under the microscope. Even scientists of some repute, conditioned to see what they expected to see instead of what was actually there, persuaded themselves and others that the sperms visible through the lenses were fully formed individuals, minute but complete. A school grew up of people who came to be called *homunculists*. They claimed to be able to see in a drop of semen tiny men and women; they even drew pictures, which can still be seen in collections of old books and manuscripts, showing an enlarged view of a minute "homunculus," sitting with head bowed, arms folded and legs crossed, enclosed in a transparent bubble which formed the head of the sperm.

The homunculist view persisted for the best part of a century, supported by the traditionalists with their predictable mixture of blind faith in their sacred writings and refusal to examine the evidence. As microscopes were improved and their power to enlarge increased, it became more and more difficult to maintain the homunculist point of view. Those who clung to it longest and most zealously were the ones who were so devout that they would not look at semen under the microscope, because the semen had been obtained by masturbation and masturbation was a sin; their spiritual descendants of the present day disapprove of artificial insemination for the same reason.

However, even the most entrenched religious faith cannot withstand reality for ever. Twenty-six years had gone by between William Harvey's statement in his *Treatise on Generation* that all living things are produced from eggs and Leeuwenhoek's first sight of sperms. Another eighty-two years were to pass before the anatomist Kasper Friedrich Wolff, in his thesis for his medical degree at Halle University, published in 1759 under the title *Theoria Generationis*, finally disposed of the whole notion of the homunculus. From his observations of the embryo of a chick through the microscope he was able to say two things: first, that the offspring is not prefabricated but is built up from "globules," and secondly that both parents contribute to its coming into being.

The word that came to be used for those "globules" of which living things are built was *cells*, taken not from the world of animals but from that of plants, having first been used by the British scientist Robert Hooke as long ago as 1665 for the microscopic cavities in cork, which he described as being made up of "little boxes, or cells." Nowadays the word cell is the accepted term for the unit of all living matter, not just that of plants, though since it comes from the Latin for a room it is really quite inappropriate to describe animal tissue, which does not show the honeycombed appearance that Hooke found in certain plant structures. However, though the word was first applied to plants it was not till the sex cells of animals had been identified and their behavior described that the fertilization of plants was fully investigated.

It took exactly one hundred and fifty years from the date when Leeuwenhoek first saw the male sperm under his microscope for another investigator to see a female egg cell for the first time. In 1827 the Estonian biologist Karl Ernst von Baer was examining the ovary of a dog. He described in his treatise *De*

Ovi Mammalium et Hominis what he found: "I could see clearly a yellowish white dot. Out of curiousity I opened one of the follicles and lifted the tiny object on the point of my knife, finding that I could see it very plainly. When I put it under the microscope I was quite amazed; I saw an ovule (an egg cell) so clearly that a blind person could hardly deny it."

That final visual proof through a dog's ovary of what Wolff had discovered sixty-eight years before by means of a chick embryo at last put the female on an equal footing with the male in the reproductive act, instead of being a mere receptacle. It could indeed be said to have been a turning point in the relationship of the sexes among human beings as well. The equality demonstrated under the microscope has no doubt played a part in convincing the male of our species to modify – if not abandon – his ancient claims to biological superiority. Much remains to be done to raise female status, but it is only a hundred and fifty years since the egg cell was first seen and already a good deal has been achieved. Predictably changes of attitude have been, and are still being, resisted by the established religious bodies, but even here there are signs that one or two people are beginning to grasp the facts of life.

The matter was put with admirable clarity by Canon Rhymes of Southwark at a meeting of the General Synod of the Church of England in November 1977. "The advances in scientific knowledge made more than a century ago," Canon Rhymes said, "are only now beginning to filter through to our moral assumptions. Before the invention of the microscope it was thought that the male 'seed' by itself was a human being in miniature. Adam was the almighty male, with the whole human race in his loins. Eve – woman – was a mere seed-bed, an incubator." Only the microscope has revealed that the roles of the sexes are complementary to each other – that humankind is created in perfect equality. "Until we have accepted that," Canon Rhymes added, "we shall never think straight about sexual ethics. So much of the Church's teaching on these matters was formulated *before* the discovery of the microscope."

As in the animal kingdom, so in the world of plants, it was not till the egg cell had been identified under the microscope that the equality of the sexes could be established. Until then, because of the general belief in male superiority, it had been assumed by naturalists since plant sex had been discovered that pollen grains contained a whole plant in miniature in the same way that human sperm was believed to contain the homunculus. Now that the whole process of plant fertilization and pregnancy had been studied in the most minute detail under the microscope, biologists realize that as soon as equality of the female is recognized in the creation of new life it follows inescapably that the female contributes more to the development of that new life and is therefore actually superior to the male.

We have seen how the male microspore, the pollen grain, germinates on the stigma and then penetrates the female tissue with a thrusting tube through which the male reproductive cells are emitted to bring about fertilization. It is now necessary to examine how the pollen grains are produced. In a typical flowering plant the pollen sac *(anther)* at the end of the stamen is made up of four lobes, each called a *microsporangium* and containing tissue which divides many times to form a large number of cells. From these develop round, well-separated *microspore mother cells*, more commonly known as *pollen mother cells*. As the anther swells, each of these mother cells divides twice to form a *tetrad* of four pollen grains *(microspores)*. During the second of these divisions the number of chromosomes is reduced to half. The process, as we have seen, is called *meiosis*, and it leaves each microspore with only one set of chromosomes instead of the two sets found in the rest of the plant cells. This double set in the ordinary cells is, as already noted, called the diploid number and usually expressed by the formula $2n$; the single set in the sex cells is called haploid and represented by n.

Before the pollen grain becomes mature the nucleus inside it divides into two, one becoming the nucleus of the "tube cell" and the other, somewhat elongated, forming a "generative cell"; this in turn will usually divide again, making two sperms, either before the pollen is shed or later when it has reached the stigma and germinated. This tiny amount of tissue within the pollen wall, including the sperms and the tube cell, represents in extremely reduced form the whole male gametophyte generation, and, as many botanists would say, the only real male sexual phase in the plant's life history.

We must now look at the female sexual apparatus to see how its development fits in with that of the male equipment we have just examined so that fertilization can take place. Within the ovary of a typical flowering plant the ovules – those structures that develop into seeds after fertilization – begin as tiny protuberances. The inside of the growing ovules, the *megasporangium*, is made up of a central mass of cells, called the *nucellus*, surrounded by one or more layers of protective cells, the *integuments*, which will eventually form the seed coat. A tiny opening known as the *micropyle*, and looking under the microscope remarkably like the entrance to a womb, is left where the outer layers do not cover the ovules completely. Like the cervix it resembles, the micropyle provides the narrow passage through which the sperm passes to fertilize the egg. At least, that is the method of entry in most cases; but there is no such thing as "normal" sex in nature, and there are some instances where the pollen tube approaches and penetrates from behind and releases its sperms into the rear end of the ovule.

Inside the nucellus a single large cell is formed called the *megaspore mother cell*. When this has grown to maturity it divides twice, forming four *megaspores*; as with the similar division of the microspore mother cell to form four pollen grains (microspores), the number of chromosomes is halved by meiosis, so that each spore has only a single set of *n* chromosomes. Unlike the pollen grains, however, the megaspores do not all survive; in most cases three of the four die in infancy, wither away and disappear. The fourth remaining one, left with no competition for food, grows rapidly into an oval structure called the *embryo sac*. While this is enlarging the nucleus divides in two, then the two divide to make four, and the four divide to make eight nuclei. As growth continues, the eight nuclei drift apart into two groups of four, one at each end, like two groups on a dancefloor waiting for partners. Then one nucleus from each group moves to the middle, where the two touch, then embrace, and finally unite to form the *fusion nucleus*.

The three nuclei left from the group at the closed end, called the *antipodals*, have little to do and soon disappear in most ovules. The three remaining at the end with the entry passage, the micropyle, form cells: the one in the center becomes the *egg cell* and the two next to it, one on each side, the *synergids*. By the time that all this development has taken place, the ovule is ripe and ready for fertilization.

For some time after this remarkably long and complicated sequence of events within the ovule had been observed under the microscope, scientists were puzzled about the function, if any, of the synergids standing on each side of the entrance. As only the egg cell was going to be sexually united with the male gamete to form a new individual, what possible purpose could be served by the formation of two surplus female cells, which seemed to have nothing to do but hang about and watch the mating of their sister? Was there some kind of ritual going on, similar to an elaborate marriage ceremony among human beings? Were the synergids playing the role of bridesmaids, and if so, why? After many observations of the process of sexual fusion between egg cell and male gamete, plant biologists believe they have found the answer. The two apparently surplus cells, it appears, are not merely spectators but play a very practical part in the process of coupling. Their function is to steer the male sex cell between them so as to guide it to their sister and make sure it does not go astray. And according to some observers they do not always perform that function properly. It is not unknown for one of the synergids instead of the egg cell to fuse with the male; the bridegroom, it might be said in such cases, has taken up with one of the bridesmaids instead of the bride.

We have already seen how after reaching the stigma the pollen grain germinates, sending out a tube which penetrates through the style and into the ovary till it comes to the entrance of the ovule. Usually the tube nucleus passes through the pollen tube first, to be followed by the two male gametes whose formation has already been discribed. When the tube has pushed its way through the micropyle into the embryo sac its tip bursts, releasing the two gametes, which are elongated and often spirally twisted, so that they appear to wriggle actively, recalling the swimming male cells that were their primitive ancestors. One of the gametes joins with the egg cell to form the zygote, which will become the seed, and later if all goes well grow to become a new plant. The other gamete unites with the fusion nucleus, which as we have seen was formed by the joining together of two of the nuclei produced during the development of the embryo sac.

Since both the male gamete and the female gamete had undergone meiosis during their formation, so that the chromosomes had been halved to the haploid number n, their sexual union will have doubled the chromosomes again to the diploid number $2n$, and the zygote will therefore have the same number of chromosomes as the ordinary body cells of its parents. The other male gamete, however, will have been outnumbered two to one when it united with the fusion nucleus, because that fusion nucleus was formed by what might be called the homosexual coupling of two haploid female nuclei, each contributing a set of chromosomes and so making the fusion nucleus diploid, with $2n$ chromosomes. The addition of the male gamete, with n chromosomes, has resulted in a threesome, two females mating with one male to form a *triploid* nucleus with $3n$ chromosomes. This becomes the *primary endosperm nucleus*, from which is formed the nutritive tissue called the *endosperm*, which fills the sac and on which the zygote feeds as it grows in the seed to become the embryo of the new plant.

After union with the male gametes, both the endosperm and the fertilized egg grow by means of the process called *mitosis*, in which the young cell divides, then the daughter cells after growing to full size divide again, then the new daughter cells grow and divide again, and so on. The resulting cells, unlike those formed during meiosis, do not undergo reduction in their chromosome number; those arising from the fertilized egg contain $2n$ chromosomes and those from the endosperm nucleus $3n$ chromosomes. The whole sequence of events during and after fertilization took many years and much painstaking research to establish. As a result it is now clear that in addition to performing the traditional role of providing a receptacle for offspring the female sex organs contribute three functional reproductive nuclei to only two from the male organ. The microscope has shown in the fullest detail that the female is not merely as important as but more important than the male. There are many variations from the typical fertilization sequence we have just examined, in particular in the chromosome number of the endosperm, but the principle of female superiority seems to remain valid in every case.

Many of the terms used to describe the reproductive organs of plants are the same as those used for animals. This is not surprising, since not only were the reproductive systems of animals – particularly the human animal – investigated before those of plants but there are some striking resemblances between the two systems. The word *placenta*, which is used for the tissues surrounding and nourishing the fetus of a mammal in the womb, which come away as the afterbirth when the baby is born, is also used for the part of the ovary wall on which the ovules are borne. The arrangement of the placenta, called the *placentation*, is used by botanists as a very important way of distinguishing between different plant families.

The ovule is attached to the placenta by a stalk called the *funiculus*, which performs just the same function as the umbilical cord does for a baby, taking nourishment from the placenta to feed the developing embryo. So close was the similarity of function that many earlier botanists used to call the funiculus the "navel-string" (as some of them called the ovary the "uterus"). To complete the resemblance to human anatomy, the funiculus breaks away when the seed is ripe, leaving a small scar – which can be clearly seen with the naked eye on large seeds such as beans – called the *hilum*. This is the exact equivalent of the scar we all carry on our bellies where the umbilical cord was once attached, and used to be called by some botanists by the same name: the navel.

When the seeds are ripe they need to be dispersed so that they arrive at a spot favorable for germination and growth. The means of dispersal are as varied as those of which we have already seen many examples in the transport of pollen to female organs, and they use the same agents: wind, water and animals. Of these, probably wind is the most frequently used. Seeds and fruits often have outgrowths which enable them to be carried a considerable distance before they fall to the ground. Such are the parachutes on the tiny fruits of Dandelions *(Taraxacum)* and Lettuce *(Lactuca)*, the plumed styles of *Clematis* and *Anemone*, and the wings of Maple *(Acer)*, Lime *(Tilia)* and Ash *(Fraxinus)*. Some seeds are so small and light that in the right conditions they are carried on the wind without needing any appendages to keep them airborne. Some pods and capsules split when they are ripe, spilling or throwing out the seed. An elegant and highly efficient distribution device is the capsule of the Poppy *(Papaver)*, which is arranged like a pepperpot with a ring of holes just below the cap; when the wind blows the stem moves, the head sways and the seeds are shaken out over a circular area.

Water may carry the seeds of many plants, such as the large coconut, with its very nearly impervious wall, and the small seeds of sedges, kept buoyant in water-resistant envelopes. Many seeds hitch a lift from animals (including human beings) by means of bristles, hooks, spines and barbs with which they cling to fur or clothes. Some seeds are voluntarily carried away because they are edible or are embedded in edible fruit. The animal may drop some, or deliberately spit out the seeds of others. Nuts are carried away by squirrels and those that are not immediately eaten are buried in the ground to provide a food store. Because of this habit, the squirrel has gained a totally false reputation as an intelligent, thrifty creature always making thoughtful provisions for tomorrow; actually it has rather a poor memory and is as likely as not to forget where it buried the nuts, so many of them, having been planted at just the right depth, will germinate and grow into new nut trees.

Some seeds are highly resistant to digestive juices and will pass right through the gut of an animal and still remain alive and capable of germination; that is why splendid crops of tomato plants are often seen growing in sewage works. Other plants do not need wind, water or animals to distribute their seeds because they have perfected devices to scatter the seeds themselves. Some of these mechanisms are explosive, such as those of many members of the pea family, like the Gorse *(Ulex europaeus)*, whose pods when ripe suddenly split and twist, throwing the seeds with considerable force in all directions. There is even the balsam called Touch-me-not *(Impatiens noli-tangere)* which at the slightest touch when the fruits are ripe will fling its seeds in a whirling motion so that they hurt like sling-shot. Perhaps the loudest bang is made by the Sandbox Tree *(Hura crepitans)*, a tropical member of the spurge family, whose seed capsule suddenly breaks up with explosive force, hurling its large seeds with such velocity that they could damage the eyesight of anyone unlucky enough to be standing near at the time.

A remarkable example of the use both of explosive devices and of animal carriers to disperse the seed has been revealed by some recent work in North Africa on species of Violet *(Viola)*. They have a two-stage method of seed distribution. The first phase, known as the "ballistic stage," is marked by the firing out of seeds with considerable force, as if from a gun, so that they are scattered at some distance around the plant. The seeds are quickly found by ants, which carry them to their nests. On the outside of the seeds are special food substances which the ants find extremely attactive and enjoyable. After the ants have gnawed off the food cells they have no further use for the seeds, so they remove them from the nest and scatter them on the ground. Very soon if the conditions are favorable the seeds germinate. The remarkable thing is that seeds that have not been gathered and gnawed by ants take much longer to germinate, that is if they germinate at all. The puzzle as to why this should be so was solved when the seeds were examined and it was discovered that in gnawing off the food substances the ants also gnaw away part of the hard seed covering, and this means that the shell is much more easily permeated by water to swell the embryo inside, which in turn is much more easily able to burst out as it germinates.

One of the strangest ways in which seeds are dispersed by animals is by means of the elephant. The plant that is said to use this method of transport is *Rafflesia arnoldi*, a parasite on the roots of trees in the jungles of Malaysia with the largest flowers in the world. It has no normal stem or leaves at all, but spends its whole life underground until its huge, brownish, foul-smelling flower bursts through the surface and rests on the ground. After being pollinated by flies and beetles it produces its soft, fleshy fruit. The problem it faces is how to get the seeds from that fruit far enough under the earth for them to reach the roots of trees on which they live. The answer is elephants. They are the only beasts sufficiently heavy to sink their feet deeply enough into the soil. They trample about on the squashy fruit and the seeds stick under their feet; by the time the pulp from the fruit has dried and the seeds are dislodged, the elephants have moved to another part of the forest, where they plant the seeds at exactly the right depth.

So far this chapter has dealt with fertilization, plant pregnancy and the dispersal of the offspring so as to give them the best chance in life. We will end with an example not just of fertilization and pregnancy but of birth control as well. It concerns the extraordinary life story of the Spanish Bayonet *(Yucca glauca)*, which grows wild in the southwestern United States and Mexico. Its hanging, bell-shaped yellowish white flowers are quite unable to pollinate themselves because of the relative positions of the anthers and the stigma, which can never touch. It is entirely dependent on a small moth, *Tegiticula yuccasella*, and the moth is entirely dependent on the yucca for food for its young. During the whole of its adult life the moth eats nothing at all. The female moth enters the flower when it opens in the evening and collects pollen from the anthers; she rolls this pollen, which has the consistency of putty, into a ball, which she carries tucked under her "chin" and to which she continually adds. When the ball of pollen is a good deal bigger than her head, she climbs onto the female sex organs of the flower, sticks her ovipositor through the wall of the ovary and lays an egg. She then climbs to the stigma and packs part of her ball of pollen into it. Next she lays another egg in the ovary, packs more of the pollen ball into the stigma, and continues with the same routine several times, till she has used up her pollen, her eggs and indeed herself; then she dies.

When the eggs hatch, the larvae in the ovary start to eat the seeds which are developing from the ovules because of the fertilization resulting from the placing by the moth of the pollen on the stigmas. Several of the seeds are eaten by the larvae, but since there are so many seeds in the ovary the loss of a few does no harm. In fact, by giving the remaining seeds more room and less competition it improves their development, and so provides a splendid example of the benefits of spacing out the offspring by practical birth control.

After gorging themselves on seed tissue, the grubs escape from their childhood home by gnawing their way through the wall of the ovary. They then spin a silken thread and lower themselves to the ground, where they pupate, protecting themselves by burying themselves well below the surface. There they remain throughout the winter. When the ground warms up they begin to stir, and in the summer they emerge as adult moths, just before the plant starts to come into flower again. The male moths spend a brief, dizzy time chasing the females and fertilizing them, and then die of exhaustion. By the time the flowers are open the female moths are ready to lay their eggs and so start the cycle all over again. In the whole field of plant sex, *Yucca* is the most remarkable example of very precise fertilization being carried out by an insect not to gather and take away food but to provide a food supply for offspring that it will never live to see.

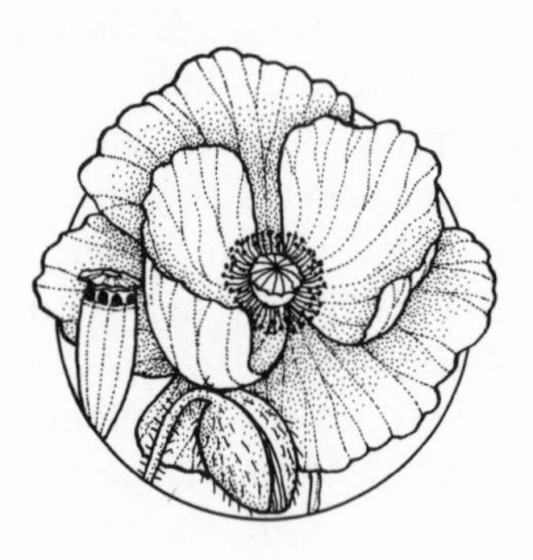

X

Virgin Birth

Though there are examples of virgin birth to be found in the animal world, such cases seem to be more or less confined to the most primitive organisms. The best known example is that universal pest of garden plants the greenfly or aphis. Female aphids give birth to a generation of young pests without being fertilized, and this seems to be very sensible arrangement when there are few if any males to mate with them. However, this virgin birth is followed by generations of sexually produced aphids as a result of the normal process of copulation; sex is a necessary part of the aphid life cycle, to keep variability going by the interchange of genes.

Many cases among higher animal organisms of what was thought to be virgin birth have turned out to be nothing of the kind. For instance, several kinds of tropical fish that bear living young, such as that favorite aquarium species the guppy, will go on producing offspring time after time in spite of the fact that there is no male fish in the tank. However, it will be found that at some time in the past copulation with a male did take place, and the result of that one sexual act is that the female carries around a "sperm bank" capable of continuing to fertilize successive batches of eggs during the rest of her lifetime.

In the higher animals there do not appear to be any authenticated cases of virgin birth that can be observed. True, the human species have developed many religions with the idea of virgin birth as a central part of their belief; however, this is not held to be a usual form of propagation but a rare – in most cases a unique – one. In any case, the belief is not biological but theological, founded on the proposition that if mankind is wicked the only way a divine being can come into our midst is through direct fertilization by a god of a human female chosen as the receptacle for his seed. Since the microscope has changed the status of women from that of receptacle to that of gamete-provider, the theory has suffered a severe blow. In biological terms we would be faced with the fact that the offspring of such a union between god and woman would not be a homozygote with $2n$ divine chromosomes but a heterozygote with n divine and n human chromosomes. Perhaps it is the death blow dealt by the microscope to the old belief that the gods came to earth by way of virgins which has led to the more modern suggestion that they came in space ships.

Virgin birth is much more common in the plant world than in the animal kingdom. In many plant species it is a normal, and in some the only, means of propagation. The botanical term for this process of reproduction without sexual union is *apomixis*, which literally means "without intercourse," and it can happen in many different ways.

To take the term in the broadest sense, apomixis may be said to range from simple vegetative propagation, where not even flowers are necessarily involved, to the full production of flowers, fruit and seeds without sexual union taking place – though in some cases a certain amount of sexual arousal and foreplay is necessary to the process. The simplest method of vegetative propagation is the spreading of a plant, as happens with many clumps in the herbaceous border of a garden, followed eventually by the decay of the middle part, so that new plants are formed at the outside of the original clump. The next simple method is the sending out of runners bearing new plants, either above ground, as in the case of the strawberry

(Fragaria), or below the surface, as with that pernicious weed Couch Grass *(Agropyron repens)*, the underground stems of which can drive their way through the hardest soil, and even through stones, sending up new plants along the entire length.

In a slightly more restricted sense, apomixis includes the production of separate organs, called *propagules*, which can be detached from the parent and kept in a state of suspended animation for some time before being planted out to grow roots and shoots from the stored tissue packed inside them and so to become new plants. The most obvious examples are bulbs such as Tulips (*Tulipa* species), which are really fat buds; corms, such as those of Crocuses (*Crocus* species), which are modified pieces of root tissue; and tubers, such as the Potato *(Solanum tuberosum)*, which are enlarged pieces of stem. Small modified bulbs, called *bulbils*, are produced by some plants, such as the Garlics (*Allium* species), and Lesser Celandine *(Ranunculus ficaria)*, a weed which is very easily spread by cultivation, the tiny bulbils being produced in large quantities and pulling themselves down below the ground when left on the surface. Several grasses, often called *viviparous* ("bearing live young"), produce heads of very small bulbils at the tops of stems; a typical species is the Viviparous Fescue *(Festuca vivipara)*. These bulbils break from the stem when they are ripe and can be carried on the wind or on the coats of animals like seeds, which they much resemble.

However, many botanists tend to restrict the use of the world apomixis to those cases where true seeds and embryos are formed without previous sexual union. Such cases are technically known as *agamospermy*, which means "making seed without marriage." This may be brought about in many different ways. The first results from the failure of the appropriate part of the ovule to undergo reduction of its number of chromosomes. The stage at which this normally occurs, meiosis, is bypassed and the chromosome number remains $2n$, so diploid embryos can develop into seeds without fertilization having taken place; such embryos cannot properly be called zygotes since they have not been formed by the coming together of male and female gametes. Sometimes all the seeds of a plant may be produced by apomixis in this way, in which cases of course all are genetically identical with the parent, not having gone through reduction of chromosome number or recombination of genes. Sometimes the same plant may produce some seed non-sexually and some sexually, meiosis and fertilization having taken place with such seeds.

As might be expected, apomixis has in some ways the same results as inbreeding, by which is meant self-fertilization, the female organs receiving pollen either from the same flower, or from a flower on the same plant, or from a plant of the same clone. The offspring in both cases resemble the parents since they share the same genetic make-up. However, the most recent investigations have shown that in fact the genetic constitution brought about by the two processes may be quite different. Self-fertilization, because it is sexual, makes interchange of genes possible during chromosome reduction and gamete fusion. After many generations of such self-mating the offspring tend to have become homozygous; that is, each chromosome of a pair resembles the other in the genes it carries. The members of each generation become uniform with each other and with the parents; incest rules out variation. However, since sex remains the method of reproduction there is always the possibility of cross-pollination by a stranger. When that happens, however rarely, variation is introduced; but generations of self-fertilization following that will once again tend to establish uniformity.

There is another possibility which is highly successful with certain species of plant: the production of some flowers which are cross-pollinated by insects, and so introduce variation, and other flowers which are self-fertilized and so keep continuity (with uniformity). One such species is the Sweet Violet *(Viola odorata)*, which bears two quite different sorts of flower. One sort is showy, scented, and attractive to bees, which eagerly visit the flowers and so bring about cross-pollination; these are the flowers which are also attractive to humans. But in addition, at certain times very small flowers are produced which remain unopened and would not be noticed by most people. These flowers, called *cleistogamous*, from the Greek for "closed marriage," fertilize themselves. The style and stamens are short and nearly touch each other; the pollen grains actually germinate while they are still inside the anthers, and when they thrust themselves out they have only a very little distance to go before they reach the stigma, push through the short style to the ovary and bring about fertilization. Such cleistogamous flowers, which are produced by several other species of plant such as the Wood Sorrel *(Oxalis acetosella)*, usually seem to occur in unfavorable conditions where there are few if any pollinating insects about, and they show the economy nature sometimes displays in its use of resources: what is the point of going to the trouble and expense of producing attractive, open flowers when there are no suitable creatures around to visit them?

Some species of plant have flowers that are not always cleistogamous but can be when necessary. They open up to receive visitors in favorable conditions but remain shut when it is cold, windy, wet or in other ways likely to cause insects to stay at home rather than venture out. The sexual organs of the flowers continue to grow and expand, however, so if the petals remain closed for long the male and female parts become pressed against each other, and if they are compatible fertilization takes place.

It is perhaps doubtful whether such self-fertilization can properly be described by the word "virgin," since sex has been involved, even though the flowers have, so to speak, copulated with themselves. Let us therefore return to the subject of apomixis, where sex does not come into it at all and so virginity is retained – if we give virginity its usual meaning of total lack of sexual experience.

The way in which the offspring of self-fertilization differ from those of apomixis is that while, as we have seen, the former tend to be homozygous, the latter tend to be heterozygous: that is, the chromosomes of each pair differ from each other in the genes they carry. They will continue to differ from each other because they are deprived of the sexual mingling that leads to uniformity in self-fertilizing flowers. There is evidence that many apomictic plants arose as hybrids between different species, and so had a mixed genetic make-up. Because of their mixed-up constitution, these hybrids were the "mules" of the plant world: quite sterile and without any hope of breeding descendants by sexual means. This hybrid nature has been demonstrated in certain species, such as the apomictic Whitebeams (*Sorbus* species), and it is thought by some botanists that most, if not all, apomicts (to give such plants their short name) originated as hybrids in the in the same way.

Because of their mixed genetic make-up, many apomicts have a very large number of what are called "microspecies": that is, descendents which, while identical with their parents (of which they are indeed a part), differ from each other in minute particulars such as slight variations of hairiness and leaf shape. Such microspecies give botanists a wonderful opportunity to discover ever new variations of a highly technical nature, and so enable them to name such a new species "*smithii*" or "*jonesii*" after themselves. Of the *obligate apomicts* (that is species which are totally impotent sexually and have to reporduce by apomixis), there are several examples among the *Compositae* (daisy family) where hundreds of different microspecies have been identified and named in this way. One such is the Dandelion (*Taraxacum*) and another the Hawkweeds (*Hieracium*).

It may be asked why, if they do not engage in sexual reproduction, such species should produce large and attractive flowers which are visited by bees and other insects. What is the point, when there is nothing to be gained sexually? The answer appears to be in part that many of these obligate apomicts have developed from species that were once sexually reproduced but in which the male organs lost their potency and became redundant. An example of this is the Lady's Mantle (*Alchemilla* species), where the stamens have become vestigial, some being lost entirely, and where if they are present the anthers are either empty of pollen or contain irregular grains which have no sexual potency left. There is, however, another suggestion that has been made as to why some of these apomicts should have flowers that attract insects. If an insect is busy on a dandelion blossom, it is neglecting other species nearby which may need its attentions for fertilization purposes. By luring the insect to itself, the dandelion is cutting down the competition by attempting to make sure that the flowers of its rival species will remain unvisited, and so wither and die without producing progeny. In this way the dandelion will be able to extend its territory by imposing sexual abstinence on its competitors.

In many cases of apomixis the process is *autonomous*: that is, the seed develops without needing any stimulus from outside. In other cases, remarkably enough, the process cannot start unless stimulation by pollen takes place; such foreplay, even though it does not lead to the complete sexual act, is necessary to initiate the production of embryos. Sometimes the stimulus seems to be merely the penetration of the style by the pollen tube; sometimes the fusion of the secondary male sperm with the endosperm nucleus appears to be needed to trigger off the process. Such a method, which begins sexually but does not go the whole way, is known as *pseudogamy*, meaning "bogus marriage."

Besides the obligate apomicts there are some *facultative apomicts*: that is, ones that can take sex or leave it alone. Cinquefoils of the *Potentilla argentea* and *Potentilla verna* groups have flowers that are still sexually functional and so able to produce the occasional new cross, which can then, if it is capable of survival, be reproduced apomictically. In this way the Cinquefoils can make the best of both worlds, sexual and nonsexual.

XI

Plant Sex and Human Sex

Long before plants were discovered to have a sex life of their own, they were used by people for sexual purposes, and they still are. The scent and sight of flowers seem to be found necessary by the human race for sexual arousal, hence their importance in fertility rites and marriage ceremonies in every kind of society, from the most primitive to the most sophisticated.

The aphrodisiac nature of flower scents has been mentioned many times in this book. Even though those scents were originally developed to attract insects, they appear to perform the same function for humans, appealing as they do to that most fundamental of sexually arousing animal senses, the sense of smell. When the first men and women appeared, it was in a world full of plants, perfuming the air with odors designed to stimulate erotic feelings. It was to the accompaniment of such scents that the earliest human sexual experiences occurred, and the same olfactory stimulus still plays a large part in promoting human sexual desire. Vast quantities of flowers are produced each year from which perfume manufacturers extract and distil the fragrant oils and essences that go into their products, claimed to arouse erotic excitement with such advertising slogans as "loves you all over."

As we have seen, many of the scents of flowers lure insects by mimicking their own sexual odors. Human beings, at any rate in industrial societies, have few if any sexual odors of their own left: the blunting of our sense of smell, the pollution of our air and the instilled sense of guilt both about sex and about body odors have robbed us of our natural source of erotic stimulus through the nose. This is of course very good for business; the same firms that profit from guilt feelings about body smells by selling deodorants also make money by selling perfumes to remedy the deficiency that their deodorants have caused. But when our own body smells have been completely removed flower scents by themselves are not potent enough for erotic arousal, so the smells made by animals for sexual stimulation are also used in the more exciting, and expensive, perfumes.

Musk, obtained from the sexual organs of musk deer and intended to make their partners feel sexy, is used for the same purpose on humans; presumably it works, which is why it is worth four times as much as gold on world markets. Other animals have their sexual odors used in the same way, such as the civet cat, which is trapped, confined to a cage scarcely bigger than its own body, and teased and aroused every few days till it ejects a strong-smelling substance from its sex glands which is scraped from beneath its tail and sold to perfume manufacturers. All these animal secretions are too powerful – and expensive – to be used by themselves, and are blended with the gentler scents of flowers to make the commercial products designed to arouse human desire and help performance.

Apart from the scents of plants, their appearance has played a very important part in human sexual customs and techniques throughout the ages. Their virtues – real or imagined – in inspiring sexual enthusiasm, or at least acquiescence, on the part of the object of one's passion have been described in herbals, medical books and works on magic since ancient times. Every culture has its store of folklore on the subject, ranging from practical advice on what plants or plant extracts a man should apply to his private parts in order to achieve or sustain an erection to how a maid may, by using the right herbs accompanied by the right incantations, cause a young man to fall madly in love with her.

One of the most important ingredients in the belief in the power of plants to assist sexual performance is the old doctrine of signatures, which held that the appearance of a plant gave a clue to its properties. Because the convolutions of a walnut resemble those of the human brain, it was believed that walnuts were a sure cure for diseases of the brain; the similarity between the leaves of lungwort and the human lung was supposed to show that they were a certain remedy for weakness of the respiratory system. Since it is not difficult to see resemblances between parts of many plants and human sexual organs, the doctrine of signatures was taken to mean that those particular plants were designed to improve sexual prowess.

One of the earliest references to this doctrine is to be found in the Bible, where in Chapter 30 of the Book of Genesis we read of how Reuben found some mandrakes in the field at the time of the wheat harvest and gave them to his mother Leah. Rachel, who was barren, was so eager to have them that in payment for them she agreed to let Leah sleep with Jacob, who though husband to them both (and father to children by both their maidservants) preferred Rachel. So potent was the magic of the mandrakes that both Leah and later Rachel became pregnant. Since then, and even up to the present day, the inhabitants of the eastern Mediterranean countries where it is native have believed it to be a sure promoter of fertility. That is not because of its medicinal qualities; indeed, it is related to the Deadly Nightshade, and its fruit if swallowed is more likely to lead to a drugged sleep – if nothing worse – than to sexual activity; its botanical name is *Mandragora officinarum*, and it is probably the plant of which Shakespeare wrote in *Othello* "Not poppy nor mandragora, nor all the drowsy syrups of the world, shall ever medicine thee to that sweet sleep which thou owed'st yesterday." The reason for the belief in the sexual potency of the mandrake is the appearance of its forked root, which looks remarkably like a human crotch, and of which the old herbalists used to assert that there were two kinds, male and female, the male having a third projection between the "legs" and the female a slit.

So strong was the fancied resemblance between the mandrake root and the human body that other resemblances were imagined too; it was held that unless it was dug up with extreme care it would scream with pain and drive the hearer mad. In the Middle Ages it was thought that to bring good luck, happiness and fertility the marriage bed on the wedding night should have a suitably sexy-looking piece of mandrake root hung at the head. The best formed roots would fetch high prices at country fairs, and unscrupulous dealers would sell pieces of quite different root, such as the common bryony, skillfully carved to look like the real thing.

Many other plants with a resemblance to sexual organs were used in the same way, to bring favorable influences to bear on the sexual act and to ensure a satisfactory and fruitful outcome. The Satyrion root, named after the ever lustful satyrs, was described by Paracelsus as "formed like the male privy parts" and was in great demand as a sexual talisman. It was a kind of orchid, and orchids were held to have specially potent properties to stimulate sexual desire, improve performance and revive flagging powers. The word orchid itself comes from the Greek *orchis*, which means a testicle, and was used as the name for this family of plants because many species have a pair of rounded tubers which strongly resemble testicles. Since the doctrine of signatures asserted that this resemblance was no accident but an indication of sexual properties, the tubers were dug up and eaten, either whole, sliced or ground, to heighten or restore virility. The mysterious way orchids appear as if from nowhere, beginning as tiny structures called *protocorms*, which arise from the association between the minute seeds and an invisible fungus, caused the old naturalists to attribute magical origins to them. They were said to arise from the semen of bulls and other animals spilled on the ground during sexual intercourse, and that suggestion added greatly to belief in their powers; what had been ejaculated by a lusty bull and then turned into a plant with testicle-like parts must confer outstanding sexual potency. It is less than a century since the extraordinary way orchids propagate themselves was discovered, and still they are thought to have almost supernatural sexual powers. A recent survey has shown that a gift of orchids is seen by women as more exciting, more dangerous and more likely to make them lose their heads than any other flowers.

Even the carrot, because of the shape of its root, was thought to bestow virility and staying power on the man who ate it; nowadays scientists tell us that it contains vitamins and other things which enable us to see better in the dark, but in earlier times their predecessors the herbalists were, as believers in the doctrine of signatures, more concerned with other things that it enabled us to do better in the dark.

Even the Stinkhorn Fungus, *Phallus impudicus*, was because of its striking resemblance to an erect penis held in superstitious awe as a symbol of sexual potency and in spite of its disgusting smell used by women to bring them fertility. It may seem absurd nowadays that merely because of shape magical powers should have been attributed to plants, but perhaps the resemblance to sexual organs gave people the chance to discuss the subject and by providing a sort of sexual "conversation piece" broke down inhibitions and led not only to talk about sex but to action.

It will be noticed that nearly all plants that by the doctrine of signatures have been held to possess sexual properties resemble male sexual organs. That may be because the male organs are more obvious and obtrusive, the female parts being hidden and secret. However, there is one strikingly female-looking exception to the rule. The biggest single-seeded fruit in the world is that of the Double Coconut, often called the Coco-de-mer, because it can float across the ocean for months, or even years, and still remain alive and able to germinate and grow into a palm tree when it reaches dry land on the other side. The giant seed, which takes ten years or more to reach maturity, looks startlingly like a woman's pelvis, both from the back and from the front, and because of its appearance it used to fetch an extremely high price as a rare aphrodisiac which was claimed to bestow on its owner sexual powers that the gods themselves might envy.

The old books are full of instructions on the use of plants as aphrodisiacs. In her charming collection of some of this ancient folklore, entitled *Gardener's Magic*, Bridget Boland calls attention to an early translation of the *Book of Secrets* attributed to Albertus Magnus, which describes how "Periwynkle when it is beate to a poudr with worms of ye earth wrapped about it and with an herbe called houslyke induces love between a man and his wife if it be used in their meals." Exactly how earthworms can be wrapped around powder Albertus does not tell us, but the last of his ingredients is still said to be esteemed by some country folk as a bringer of married bliss. Its modern spelling is Houseleek, its botanical name is *Sempervivum tectorum* and it has the old fashioned popular name "Welcome-home-husband-however-drunk-you-be" because it is said that in a home with a houseleek on the roof the man is always capable of an adequate sexual performance whatever his condition may be.

Many other plants were believed to stir up desire or improve performance. The old herbalist Gerard recommended the fleshy corms of the Sowbread *(Cyclamen)* for the purpose: "Being beaten and made up into flat cakes, it is reported to be a good amorous medicine to make one in love, if it be inwardly taken." The old herbals are full of recipes for dishes in which erogenous plants were secretly included to be given to the object of one's desires "for to heat up the amatory parts." Many of these aphrodisiac ingredients may well have included substances that a modern pharmacologist would recognize as having a stimulant, or even an irritant, effect on the sexual organs. Others were considered potent only because of the plant's sexy appearance. The phallic look of asparagus gave it high prestige among aphrodisiacs, so that Culpeper, writing in the seventeenth century, asserted that if it was boiled in wine and taken in the morning it "stirreth up bodily lust in man or woman," and the White Deadnettle *(Lamium album)* was credited with extraordinary erotic powers because the stamens in the flower lie side by side like a couple in bed.

To invite Venus, the goddess of love, into the bedroom, the floor should first be strewn with rushes and then over those sprinkled with sweet-smelling herbs such as Thyme and Marjoram, Woodruff and Meadowsweet, with Mint, Vervain and Valerian, and, in spring, with Violets. When wild roses were blooming in the hedgerows a handful of their petals freshly gathered and placed beneath the bed could be relied upon to make a woman willing, and some flowers of broom – sacred to Mars – to render a man potent and eager. The bed itself should, naturally, be strewn with Lady's Bedstraw *(Galium verum)*, the stems of which as they dry and darken give off a delicious smell of new-mown hay, caused by what we now know as a chemical called coumarin, which, to judge from what goes on between lovers in haystacks, seems to have a genuinely aphrodisiac effect.

The sheets, so the old herbals say, should be scented with Marjoram in honor of Venus, and the pillow stuffed with Verbena – though its smell is so pungent that Bridget Boland tells us that if we want to try its effects in a present-day bedroom a single sprig thrust among the down of a modern pillow might well be enough. Finally, it was strongly recommended that a large bowl of potpourri containing certain fragrant herbs and spices should be placed on a bedroom table; if stirred from time to time with the wedding-ring finger (though whether this needs to carry a wedding ring or not we are not told) the contents of the bowl will release their scents and so "induce the mood of love."

In the days before contraceptives had been developed, and such crude methods as were available were not only ineffective but in many cases declared illegal, or sinful, or both, there was always the danger that any aphrodisiac that was successful would also increase the risk of unwanted pregnancy by increasing sexual activity. Many recipes for potions to arouse desire or improve performance would therefore include ingredients intended to avoid the consequences. Most of these forerunners of the pill were both harmless and useless, and led to a brisk trade in much more profitable concoctions designed to bring about abortions and containing such dangerous substances as plant material infected with ergot, sometimes used with fatal results. However, in spite of the occasional accident caused by ignorance or cupidity, human sex seems to have benefited through knowledge of the properties of plants. Maybe in a large number of cases those properties were magical rather than real, but if belief, however irrational, in the virtues of plants gave lovers confidence in their powers, no doubt that improved sexual performance for what we would now call psychological reasons.

In addition to the charms and potions that were intended to arouse desire, there were others whose purpose was to quell lustful feelings. These counter-aphrodisiacs were much sought after by men going off to the wars and crusades and leaving their wives and mistresses behind. It was all very well to lock them into chastity belts to prevent them from straying, but how much better (and cheaper) to destroy their sexual desire by means of a herbal anti-love potion. Amazing mixtures were made up for the purpose, compounded of such things as henna flowers, mallows and bad eggs blended with onions; it may be that such concoctions were remarkably effective sexual turn-offs simply because they were so disgusting. Celibate religious communities were constantly on the lookout for ways of reducing sexual desire, so monasteries and nunneries built up elaborate sets of instructions for the manufacture of counter-aphrodisiacs. The twelfth-century nun Hildegard of Bingen, as obsessed with the evils of the mandrake root as present-day purity crusaders are with the wickedness of modern sex aids, compiled a list of potions to reduce desire and ward off carnal thoughts, compounded of such things as shoots of broom and roots of cranesbill mashed into a paste and spread on the body.

At the present day the use of plants either to arouse or to reduce desire seems to have suffered a decline, except in the case of perfumes, which have become a huge and highly profitable industry. The use of flowers in courting, at betrothals and in marriage ceremonies seems to be traditional in every type of culture, perhaps in unconscious acknowledgement of the fact that sex was invented by the plant world.

The language of flowers, by which lovers were able to send each other messages of tenderness, rejection, grief and hope, has a long history and an elaborate set of rules. It seems to have come in for something of a revival recently; feature articles in magazines, and even whole books, are devoted to giving readers such useful information as that red roses mean "I love you," love-in-a-mist says "I am perplexed," raspberry (somewhat surprisingly) declares "I am sorry" and basil means quite simply "I hate you."

The present trend, in these days of apartment dwelling, for

people to grow flowers and plants in their homes is adding one more chapter to the history of the age-old association between humans and plants. According to a recent survey in the United Kingdom sponsored jointly by the Flowers and Plants Council and the Pot Plant Growers' Association, the growing of plants in the home restores a feeling of companionship to the lonely. Wives who have lost their husbands and husbands who have lost their wives see plants not only as someone to share their lives with but as a love object; they do not stop at talking to their plants but fondle and caress them as well, and they swear that the plants respond to their advances. As for married couples living together, they find that flowers and plants make for a happier relationship in every room in the house except one: the bedroom. The husband, it appears, sees a plant not only as a person but as a sexual rival. To quote from the London Daily Mirror's report on the survey: "What's green, pretty and frightens husbands? Men, it seems, feel threatened in bedrooms . . . by potted plants. Men actually resent the plants as being too intrusive."

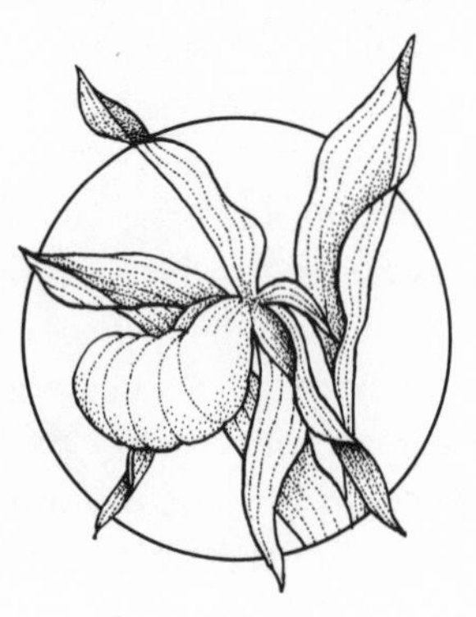

XII
The Future of Plant Sex

The use of such sex aids as electrical vibrators to improve an unsatisfactory sex life has proved as much of a blessing to plants as it has to people, and for the same reason: to make up for the lack of partners, or for their inadequacy.

Many crops grown in glasshouses are denied normal gratification of their sexual needs because they lead an artificially sheltered life, deprived of visits by insects and wind, which cannot therefore perform their natural function of bringing about pollination. Even flowers that are perfectly capable of self-fertilization are not always able to get the pollen from the male to the female organs unaided. Indeed, that is why commercial growers of flowers under glass go to great lengths to keep out insects, because once a flower has been pollinated it quickly withers and fades, concentrating on producing offspring instead of trying to keep its looks.

Growers of tomatoes have precisely the opposite problem. They want the flowers to be fertilized and to fade as quickly as possible, so that the fruit may develop without delay. Noticing how many flowers on greenhouse tomato plants often fail to set fruit satisfactorily, growers have long tried to find ways of improving performance. Manipulation was found to achieve quite good results, and it became common practice to go round the plants shaking them vigorously by hand to encourage the emission of pollen. That same hand method is still used in many places, but it is tiring and does not always bring results. The development of the electrical vibrator – actually known as the "artificial bee" – has improved things greatly; the latest models, properly used, give very satisfactory results. For a time after its invention and first enthusiastic reception, the vibrator fell out of favor because it was said to spread disease, but with improved techniques it is becoming popular again.

No doubt we shall see many more sex aids invented for use on plants in the future, and later in this chapter we will deal with other ways in which the influence of humans, intentionally and otherwise, is altering the sex life of plants. But first let us look at some of the directions in which plants appear to be evolving by themselves, regardless of the ways in which we exploit, manipulate and modify them. A most remarkable and exciting example of what may be the evolution of a new species under our very eyes has been found during studies carried out by R. M. Straw on the genus *Penstemon*, a member of the figwort family known to gardeners as Beard Tongue. There are about one hundred and fifty species of *Penstemon* native to the United States and Mexico, and the size, shape and color of the flowers vary greatly from one species to another according to the tastes and habits of the different pollinators they attract. One of the most spectacular is *P. centranthifolius*, which has flowers of a dazzling carmine, an irresistible color to the hummingbirds which visit them. The flowers are narrow and tubular, to suit the birds' long beaks, and are without any lip to form a landing place, since the birds hover in the air while helping themselves to nectar.

A very different species is *P. palmeri*, which has creamy white flowers, tinged pink, with a very wide throat and an open mouth to accommodate the large bees that visit them and are able to crawl right inside. The lower lip is broad and strong, to form a good landing place, and the flower stalk is thick and powerful, to bear the weight of the heavy visitors. While investigating relationships within the *Penstemon* genus, Straw was struck by a third species, *P. spectabilis*, which seemed to be in many ways intermediate between the other two: its flowers were purplish lilac, with a deep throat that was neither narrow enough to be designed for the beak of the hummingbird nor wide enough for a large bee to be able to force its way inside. This third species strongly resembles a hybrid made by crossing the other two together, but the remarkable thing is that instead of being visited by hummingbirds – for which the color and the way the blossoms are held are both unsuitable – or by bees – which cannot get in – the flowers are visited by a kind of wasp, which, being smaller than the bees, can crawl comfortably inside and which has become the regular pollinating partner of *P. spectabilis*.

The most interesting aspect of the affair from the point of view of evolution is this: if hummingbirds visited the hybrid they would tend to cross it back to the red, narrow flowered parent, which they prefer, and if bees visited it they would tend to cross it back to the white, wide-flowered parent, which is their regular floral partner. In either case, the new qualities of the hybrid offspring would die out and be lost, and the old established characteristics of the parents would reassert themselves. By happy chance, the finding by the hybrid of a new pollinator in the shape of a wasp, which is attracted by the offspring but neither cares for nor is able to cope with the parents, ensures that interbreeding occurs only between the progeny, not between them and the parents. The result will be that the best features of the hybrid will be preserved and strengthened by natural selection and the worst features will be eliminated, so that within several generations a new species will have evolved, with uniform genetic characteristics. Any modifications that make the flowers more attractive to the wasps will be retained and improved; anything that turns the wasps off will be bred out and lost. If the process is successful, it may be that the wasps will become so wedded to the new species that they forsake the old ones which they used to visit before. The ideal from the plant's point of view is that it should provide so much satisfaction to a particular pollinator that that pollinator never feels the need to seek satisfaction elsewhere. Such faithfulness of one species of visitor to one species of plant is called by scientists *monotropy*, which means roughly "driven in one direction," and has during the course of evolution given those plants that have achieved such a relationship the advantage of not having their time wasted by carriers of unsuitable pollen. A monotropic creature is one that is totally incapable – for physical, physiological or other inbuilt reasons – of making use of any kind of plant but one. It is completely bound to that plant and unable to transfer its attentions to any other.

Other species of pollinator have a certain amount of individual freedom and can transfer their attentions from one species of flower to another as the fancy takes them. Some are *polytropic*: that is, they are able to visit many different kinds of plants with quite dissimilar flowers. Others are *oligotropic*: that is, they are restricted to certain groups of related plants. It will be evident, then, that the pollinator of a new plant, such as the wasp that started to visit the new hybrid which became *Penstemon spectabilis*, could not be monotropic, or it would have been unable to experiment with anything in the least different from its regular floral partner. It would have to be either polytropic or oligotropic. From the plant's point of view the latter would probably be preferable, since it is easier to make a faithful partner out of a visitor with restricted tastes and habits than out of one that is quite fancy free.

As insect that has the ability to exercise a certain individual freedom of choice may decide to use that freedom by voluntarily becoming faithful to one species of plant, whose flowers, for some physical or psychological reason, it finds to its liking. Sometimes the insect will have tried several different types before it decides on the one it wants to settle down with. In many other cases it will never exercise the choice which is its theoretical right but will establish a lifelong faithful relationship to the first species of plant with which it has relations.

On its maiden flight when it first leaves the nest a young virgin insect has a lot to learn. It is driven by the instinct to possess a flower, but its first unpracticed attempts are bound to be somewhat clumsy. Only the superficial attractions of color, shape and smell will lead it to try its luck with one particular flower rather than another on that first occasion. Later, as it gains practice and learns how to get what it wants from the flower, it ignores those superficial attractions. Experiments with bees have show that once they have gained experience they will still visit the blossom of their choice even if its petals have been removed and its scent masked by some other smell; they have learned to disregard appearances and settle for solid satisfaction of their desires. If a virgin insect has a more or less satisfactory experience with the first flower it tries near home, on that maiden outing, it will gain confidence, lose its awkwardness and soon acquire the skill to get what it wants with the least fuss and bother. It will, so to speak, have settled for the girl next door and given up any intention of seeking satisfactions further afield.

In either case, whether the insect has had several adventures or none before it decides to become faithful to one flower, forsaking all others for as long as it lives, the scientists call its behavior, not surprisingly, *constancy*. The essence of constancy is that it should be – or at least appear to be – a matter of free choice on the part of the pollinator. For those cases where it is not clear whether faithfulness is monotropic – that is, dictated by membership of a particular species – or forced on an otherwise fancy-free creature because no other flowers are available for it to visit, the scientists use the word *fidelity* to describe its behavior. In the case of the wasp that has become the regular partner of *Penstemon spectabilis* we do not know the degree, if any, of free choice, so fidelity would seem to be the most appropriate term.

The creation of new species in this way, from hybrids which find their own pollinators, is a process that is being increasingly studied. Already another species of *Penstemon* has been found which is thought to have developed in the same way as that described for *P. spectabilis*, but later. This time, it seems, the wasp-pollinated species was in its turn crossed with the bird-pollinated species, and among the progeny was one that was not attractive to or suitable for the wasps, the birds or the large bees, but exactly right for a species of small bee which has become its faithful pollinator. No doubt many more examples will be found of this highly evolved constancy between one flower and one visitor. As we have seen, it has its advantages in increased efficiency, but it also has its dangers. If a plant has become totally dependent on one insect and that insect is for some reason or other not available, the plant will not be able to reproduce sexually, unless like the Bee Orchid it can manage to fertilize itself if no visitors come to attend to its needs. In many other cases where self-fertilization is not possible, either because the sexual organs are unable to touch each other or because of incompatibility, the plant will die out through sheer sexual frustration.

The danger of this happening has been greatly increased by modern man and his use of powerful insecticides, many of which are quite indiscriminate in their effects and just as likely to kill harmless pollinators as to get rid of harmful pests. Already many of the beautiful wild flowers that used to abound in and around cornfields have vanished or are in grave danger of becoming extinct because of the spraying of crops with deadly chemicals designed to kill insects. In addition there is increasing use of herbicides, designed to kill plants directly by applying growth hormones in such strength that instead of promoting healthy growth they cause monstrous cell development which cripples the plants. Unfortunately all these sprays tend to destroy many of the most beautiful species, because they are the ones which rely most heavily on sex, the appeal of their flowers being, like all forms of beauty, based upon sexual attraction. The plants that most readily survive the poisonous chemicals, whether insecticides or herbicides, are the least attractive ones which either reproduce themselves without sex or are promiscuous, relying for pollination on a variety of insects which manage to survive the sprays.

The efforts by man to control and manipulate plants for his own ends are transforming the worlds of agriculture and horticulture, and in many cases are tending to cut out or restrict sexual reproduction. Because sex creates variety, it is found undesirable by the growers of many commercial crops, who want to produce an absolutely standard product, without variation, suitable for mass marketing. In some cases they do this by eliminating sex altogether and in other cases by controlling it strictly by methods similar to that of artificial insemination.

An extreme example of the first method is the technique of

tissue culture. As we have seen, all living things start as a single cell, which in higher organisms then divides again and again, building itself into an assembly of what may amount to many millions of cells, each becoming specialized for a particular function. If any cell can be taken while it is still young, before it has started to specialize, and placed in the right conditions, it should be possible for it to multiply itself and eventually to grow into a complete new organism. And so it has proved to be. Any plant with particularly good qualities can now be propagated non-sexually to produce millions of absolutely identical replicas of itself. Until a few years ago an unusually desirable orchid, especially one that had been awarded a gold medal or a first class certificate, would cost some rich collector a fortune, because it would take many years to produce a new plant by natural growth. Now it is possible to mass-produce exact copies of that same orchid by laboratory methods quite cheaply. Under a microscope, and in strictly controlled sterile conditions, a minute portion of tissue, consisting of a small number of cells, is cut from the growing point of a plant with a surgeon's scalpel. This piece of tissue is placed on nutrient jelly, containing a mixture of chemicals, in a test tube or flask and put under artificial light. Within a short time the tiny group of cells will multiply and turn green. If the test tube remains stationary, the effects of light and the force of gravity will cause that mass of cells to differentiate, those at the top beginning to form a shoot and those at the base sending out roots into the jelly. If that is allowed to happen, only one new plant will be produced. The technique that has been developed by commercial growers is therefore to keep the test tube continually on the move. It is fixed to a revolving wheel which is kept turning by an electric motor at a constant rate. Since the mass of cells has no way of knowing which way is up and which is down, it cannot form a base or an apex and so cannot differentiate into shoots and roots. Instead, all it can do is to go on growing bigger and bigger. After this has continued for several weeks the test tube is opened and the, by now, bloated mass of cells is cut up into tiny pieces, each of which is placed on fresh nutrient jelly in a separate test tube, and so the process is continued. In this way hundreds, or thousands, or even millions of new portions of growing tissue can be rapidly produced from one original plant. When there are enough, they are transplanted into flasks or test tubes again, only this time they are not kept revolving but allowed to stay still. The effects of gravity are felt and the clusters of cells begin to differentiate into shoots growing upwards and roots downwards; soon leaves appear, and within a few months a batch of identical, mass-produced plants will have come off the assembly line and be ready for sale.

There is no reason why this same technique of clonal propagation could not be used to increase the numbers of any living organism, animal as well as plant. Already experiments are well advanced which one day may lead to the mass-production by tissue culture of unlimited numbers of identical cows, surgically produced from a specimen with a particularly high milk yield, or millions of hens from one record-breaking egg layer, with no need for the sexual services of the cock. In theory, it should be perfectly possible to manufacture human beings in the same way, by removing a few cells from a person selected for some particular quality and growing those cells in artificial wombs in carefully controlled conditions. A specially orthodox individual, bright enough to understand an order and efficient enough to carry it out, but too stupid and unimaginative to question it, could be clonally propagated to provide a dictator with an army of identical troops, totally obedient and completely without fear of death, since the destruction of one unit would be of no importance, the survivors being parts of the same organism. Sex, and with it variety, will have gone; sexlessness, and with it uniformity, will have taken over.

But that is a dream – or perhaps a nightmare – of the future, and is in any case beyond the scope of this book. In the world of plants, with which we are concerned, already the technique of tissue culture is being used on many crops of economic importance as food. Asparagus is an interesting example. Until very recently all asparagus plants were raised sexually from seed, and the seedlings were very variable. To complicate matters the male and female plants are separate. Perhaps because of the phallic appearance of asparagus shoots (which gives them a quite unwarranted reputation as an aphrodisiac) it has long been asserted that the male plant is superior to the female. The American *Standard Cyclopedia of Horticulture* refers to the males as being "more vigorous and more productive of good shoots" than the females, and goes on to mention the difficulty of tearing out "the less desirable female plants." The Royal Horticultural Society's authoritative *Dictionary of Gardening* is even more male chauvinistic; if refers to the males as "more thickset and denser" and the females as "weaker," and recommends that the asparagus bed should be restricted to males only; though it does later suggest that a few females might be grown in a bed of their own for breeding purposes only. The evidence for this male superiority seems to be strangely lacking; it appears to be a tradition derived from the ancient doctrine of signatures and to have little if any foundation in fact; it is a simple observation that female asparagus shoots are capable of being every bit as phallic as male ones. And now with tissue culture techniques the most desirable specimens are being commercially multiplied from plants of either sex, irrespective of prejudices about male superiority.

So much for artificial reproduction of plants by non-sexual means. Another method of manipulation by strictly controlled sexual means, similar to the technique of artificial insemination with domesticated animals, is the production of what are called F1 hybrids. The term F1 means "first filial generation," and refers to the raising of hybrid seed by crossing together two pure strains of a plant and sowing that seed to produce plants of very much improved growth and performance, showing both hybrid vigor and extreme uniformity. The way in which these F1 hybrids are created is to grow the parent strains, which have been inbred for many generations, in complete isolation from each other, and then when they are sexually mature to mate them forcibly with each other, often by hand pollination. This "arranged marriage" between the two inbred strains results in robust mongrels, far more vigorous than the parents, but these mongrels are not permitted to breed at all, and so never have offspring of their own. The pure lines continue to be inbred and kept in isolation, and the arranged marriage between them is performed every season.

Because of the extra cost involved in maintaining the genetic purity of the inbred parental lines, F1 hybrid seed is more expensive than ordinary seed. What adds most to the cost, however, is that if pollination of one parent by the other is carried out by hand the process is very expensive compared with the free services of insects. To increase the quantity of seed produced and bring down the cost, large scale sowings of the two parental lines are being made in the same field and the pollination left to insects, particularly bees; hives are often brought into the field for the purpose. To produce hybrid seed, crosses must be made *between* the two parental lines, so pollination *within* the same line is to be avoided, and for that reason self-incompatible parental strains are chosen. One crop in which this technique is being used is Brussels sprouts, which give far better and more uniform results as F1 hybrids. An unexpected problem has cropped up, however: bees can distinguish subtle differences between the parent lines and show a maddening faithfulness, so instead of crossing the two lines they tend to stick to the flowers of one or the other. Perhaps the solution will be to breed and release blowflies, which, being less fastidious, make no distinction between the two lines and show no tendency towards faithfulness.

To produce hybrids of this type, the parents have to be compatible when they are mated – otherwise they would not produce any offspring. Where the parents are incompatible but it appears that if they could in some way be mated the offspring would be profitable, new techniques are being used to try to break down the antagonism between them. Sometimes an artificially created environment will bring about compatibility which in natural conditions could never be achieved. In some cases incompatibility will break down after the flowers have been open for some time, so that offspring can be produced by mating elderly flowers which in their youth would have nothing to do with each other. Sometimes the reverse is the case; the antagonistic chemical substances do not begin to appear until flowering has started, and in these cases bud pollination before the flower opens may be successful. Hormones too may help to break down resistance.

In some cases plant breeders have been able to change the sexual characteristics of the progeny of these arranged matings, so that male or female offspring can be produced at will. One of the most important examples of such advanced genetic engineering is that of the cucumber. Nobody has been able to find cucumbers growing in the wild, so it is uncertain where they originated. All that is known is that they have been cultivated from ancient times, and during the course of that long association the most highly esteemed varieties have become *parthenocarpic*: that is, they are able to produce fruit without fertilization. Such fruit is sweet and juicy and contains no seed. The male and female flowers are quite separate and distinct, though they appear on the same plant, the male producing masses of golden pollen and the female carrying a tiny fruit which lengthens and swells whether sexual union has taken place or not. The males are nothing but a nuisance. If they are allowed to fertilize the females the fruit is hard and bitter; the flesh becomes tough and full of seeds. Until recently, therefore, growers of cucumbers had to take precautions against unwanted pregnancy by keeping out insects and wind that could have brought the male pollen to the female flowers; if that was not possible they had to go round the plants every day nipping off the males before they had time to open and threaten the females with fertilization. Now breeders have overcome the problem by developing F1 hybrid cucumbers that produce only female flowers. There are no males to interfere with them and hardness and bitterness are things of the past.

So far we have dealt with deliberate interference with plants by people for their own ends. Often, though, the activities of man have quite unintended effects when he destroys living relationships built up through the ages.

Perhaps the most extreme example is that of the Dodo and the tree whose partner it used to be. *Calvaria major* is probably the rarest tree in the world. It grows only on the island of Mauritius in the Indian Ocean. The strangest native of Mauritius used to be the Dodo, an odd-looking, ungainly bird the size of a swan. Because it was a peaceable, slow-moving creature with no weapons of defense, it was mercilessly hunted by human invaders of its territory. Since the flesh of the Dodo was so uneatable that the Dutch explorers called them walgvogels, which means nauseous birds, this hunting was done not for food but in the name of "sport."

The last Dodo was hunted to death in 1681, since when the species has been extinct. Ever since that date, no seeds of the Calvaria have been known to germinate. Only a dozen or so elderly specimens survive. Since all of these existed before the Dodo was wiped out, even the youngest must be at least 300 years old. For many years, naturalists have puzzled over the nature of the relationship between the Calvaria tree and the Dodo. The ancient trees still continued to produce fruit, but for some reason the seeds did not seem to be fertile. Since the Dodo had never been seen to fly, how could it ever have fertilized the flowers? It could hardly have climbed the tree; it was much too heavy and clumsy. Unless somebody found the answer to the riddle, it would be impossible to produce new trees to replace the few survivors now on their last legs, and soon the Calvaria would be as extinct as the Dodo itself.

The riddle has recently been solved. Dr. Stanley Temple, of the University of Wisconsin, is reported to have found that the seeds still being produced are not infertile after all. The part the Dodo used to play was not at the stage of fertilization but later on in the reproductive process. It was not a mating partner, but a sort of midwife. The trouble was not with the seed, but with the thick wall of the fruit that surrounded it; this was so tough that the embryo inside was quite unable to break through it. Birth had to be in some way induced, otherwise the embryo would, after a time, shrivel and die inside its stone hard "womb."

As a result of his observations, Dr. Temple formed the view that during the course of its evolution Calvaria had become dependent on the Dodo to grind down the walls of the stones until they were thin enough to allow the embryo to break through when it reached the right stage of development, just as a chicken breaks through an eggshell. When the fruits dropped from the tree, the Dodo would pick them up from the ground in its large beak and swallow them. The stones would remain in the bird's powerful gizzard perhaps for several days, during which time they would be continuously ground so that the shell became considerably thinned and weakened. It must have been a hazardous process, because the Dodo's gizzard was probably strong enough to crush many of the stones, after which the contents would pass into the bird's digestive system and be assimilated. However, Dr. Temple argued that a proportion of the *Calvaria* stones would be strong enough to resist being crushed. Their spherical shape would be highly efficient at enabling them to withstand the pressure. These stones would go through the bird's intestines and eventually be passed with its droppings: during the process they would have become weakened enough to germinate.

To test his theory, Dr. Temple fed fresh Calvaria stones to turkeys, which he thought were perhaps the nearest living things to the Dodo in terms of size and gizzard power. The outcome of Dr. Temple's experiment, as reported in the journal *Science*, was highly encouraging. The stones were retained in the turkeys' gizzards for a considerable time, in some cases as long as six days. Out of a total of 17 stones, only 7 were assimilated by the birds. The remaining ten were either vomited up or passed out in the birds' droppings. When he examined the stones, Dr. Temple found that, as he had hoped, their walls had been ground down. He planted all ten, and three germinated. So perhaps the Calvaria tree has a future after all, thanks to the turkey "foster parents."

XIII
Sex Problems

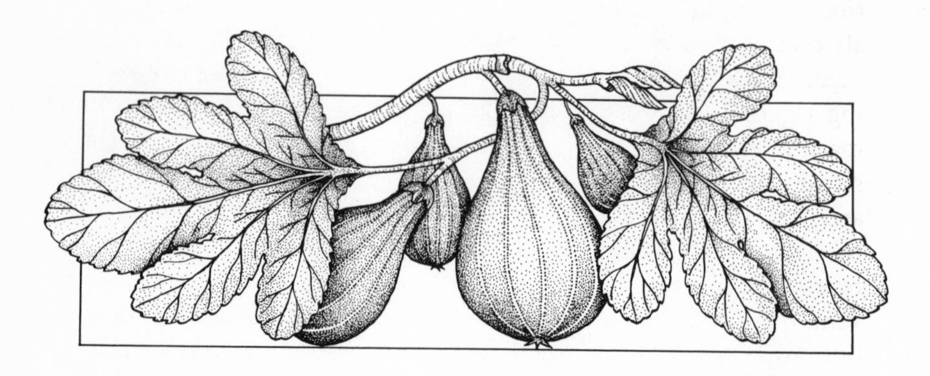

The main sexual problems that plants have to face are premature ejaculation, frigidity and unfaithfulness. Premature ejaculation means just what it says: sexual emission by the male organs before the female ones are ready. The very invention of sex created the problem, and so made necessary the development of many of the mechanisms and techniques described in these pages in order to overcome it. Nonsexual reproduction does not face the same difficulty; if there are no sexes involved, one does not have to wait for the other, and spores can grow directly into new plants whenever they fall in the right place, without having to go through the stages of mating and fertilization.

For sexually propagated seed plants life is not so easy. The male microspore, the pollen grain, must not only reach exactly the right part of the female sexual organs, the stigma (which may be only a millimeter or two across) but must do so at precisely the right time, when that stigma is sexually receptive. If it is too soon or too late, fertilization will not take place and the pollen will be wasted. As we have seen, in the normal unfolding of a flower the male parts develop before the female, which is why protandry is so common, with pollen being shed before the stigma is ready. That is all very well if there are other flowers nearby which have passed through the male phase and reached the female stage; in fact that will ensure cross-pollination, with all its benefits. But even then there must be something – wind, water, insects or other creatures – to bring the pollen to the stigma, otherwise mating will not occur. For that reason methods have evolved both to delay emission till conditions are right and to enable the male organs not to shoot out all their pollen at once but to hold some back so that they are capable of more than one sexual performance instead of exhausting themselves in one go. When men have the problem of premature ejaculation it is because their genital organs are too sensitive, and so they are recommended to relax, to take it easy, and perhaps to apply one of the creams or sprays now available to reduce sensation and so delay orgasm. With flowers, the reverse process often takes place: to conserve the pollen and prevent if from being wasted, the male organs are made not less but more sensitive. A good example is the Cornflower *(Centaurea cyanus)*, which has beautiful flower heads with large, bright blue outer florets and smaller, somewhat purplish inner ones. The outer florets are quite sterile, with abortive sexual organs, and only exist as advertisements to attract the attention of visitors. The inner florets are bisexual and promiscuous, drawing a mixed assortment of insects, particularly flies and bees. The male organs containing the pollen are joined together into a dark purple anther tube with a beak at the tip, standing erect and protruding above the rest. Unlike that of the human male member, it is not the head that is sensitive but the shaft beneath, which is made up of separate curved filaments. These filaments are extremely ticklish and when stimulated they suddenly straighten and contract, shortening by several millimetres and drawing the anther tube down with them. Inside the tube, and surrounded at its base by the sensitive filaments, is the style, which carries a little way beneath its tip a ring of stiff hairs rather like a chimney sweep's brush.

The way the mechanism works is this: a bee or other insect, attracted by the bright but sexless outer florets and finding they have nothing to offer, will wander on to the less conspicuous inner florets and discover that they contain nectar at the base. To get at the nectar the bee will have to stick its tongue down through the floret and in doing so is sure to touch one or more of the ticklish filaments, which will instantly contract. The anther tube will move down and the bristles on the style will brush out a quantity of pollen from inside it and plaster it on the bee's body. The stigma on the style in the floret will still be in a virgin state, its two lobes tightly closed together. If the bee then goes on to another floret that is old enough for the two lobes of its stigma to have opened out and become sexually receptive, cross-pollination will take place.

After some time has gone by since they were stimulated the filaments begin to curve again and elongate to their original length, the anther tube covering the tip of the style again like the foreskin on an uncircumcised penis after detumescence. The performance can be repeated several times if another visitor comes along to stimulate the male organs, though the amount of pollen emitted will never be quite as much as it was the first time and it will get less on each occasion till there is no more to come. Then the floret, exhausted of pollen and finished as a male, will change into a female, the style growing longer and protruding through the anther tube, and the lobes of the stigma parting and giving out the necessary secretions to welcome pollen from other florets.

Many other families contain species which use various means to prevent pollen from being wasted when there are no pollinating agents around. In some cases these means are not quiet ways of dispensing pollen in measured amounts over a period but violent methods using a good deal of force. *Pilea muscosa*, a member of the nettle family *(Urticaceae)*, is called the Gunpowder Plant, the Pistol Plant or the Artillery Plant because as soon as it is ready it will puff out clouds of pollen at the slightest touch. Several members of the pea family, such as the valuable fodder crop Alfalfa or Lucerne *(Medicago sativa)*, have spirally coiled sexual organs which develop such stresses within them as they grow that when an insect pushes its way into the flower the stresses are released with a jerk and pollen is thrown all over the visitor. The Trigger Plant *(Stylidium graminifolium)* carries spikes of pink flowers containing a column in which the male and female sexual organs are joined; the "trigger" device is so finely adjusted that as soon as it is touched it fires pollen at the visitor with great speed and accuracy.

Sometimes in spite of all precautions against premature ejaculation some hermaphrodite flowers are left in the female stage, sexually receptive and waiting to be mated, after the male organs have lost all their pollen and withered and dried up. That often happens at the end of the flowering season among such families as the *Umbelliferae*, in many species of which the anthers actually drop off after the emission of their last grains of pollen. Other members of the family, such as Sweet Cicely *(Myrrhis odorata)*, have overcome the problem by producing some entirely male flowers at the end of the flowering period, purely to provide a fresh lot of pollen to satisfy the sexual requirements of those last female flowers and prevent them from dying unmated.

All these methods of keeping pollen from being prematurely discharged before a suitable visitor appears create another problem: the longer the pollen is held before being released the more carefully it has to be protected and preserved. As we have just seen in the case of the Cornflower, hiding the pollen and dispensing it a little at a time is a good way of preventing waste, but concealment is also a method of pretecting against thieves which are always trying to steal pollen without carrying out any pollination in return. Hiding the pollen in a safe place also protects it against being snatched away by the wind, a great hazard for insect-pollinated plants. Most important of all, pollen needs protection from moisture, which can quickly kill it.

Some flowers solve the problem neatly by closing themselves up at the approach of bad weather. That has the advantage not only of keeping out the wet but of making sure that the pollen (and for that matter the nectar if there is any) is not exposed unnecessarily when the appropriate pollinator is not active. And that brings us to the second of the main sexual problems of plants, the problem of frigidity. For satisfactory sexual relations, not only do the male and female organs need to be ready at the same time, but they also need to be at the right temperature if the union is to be successful. That is why in cool climates few flowers appear during the winter; not only is there a lack of pollinating insects in the cold weather, but freezing temperatures are as unlikely to lead to sex among plants as among people. For that reason in most cases even the earliest flowers do not come out till the Spring is well advanced and the danger of severe frost is over.

The third type of sexual problem for flowering plants is that of unfaithfulness. As we have seen, many flowers are promiscuous, open to any and every type of visitor that happens to come along and not particular about which they allow to pollinate them. The most primitive flowers belong to this type; they are open and bowl-shaped and they do not attempt to hide their attractions; the pollen, and nectar (if any) which they have to offer are displayed for all to see, and a large proportion is taken by primitive creatures such as bugs and beetles which do not pay for the goods they have helped themselves to by bringing about pollination. The trouble with such flowers is not that the visitors bring no pollen with them but that the pollen with which they are dusted from the previous flower they visited is likely to be useless. That is because it probably comes from a quite different type of plant and so is incapable of performing cross-fertilization; no species can be successfully pollinated except by pollen from its own species or a closely related species.

Since contamination by foreign pollen is extremely inefficient, not only because fertilization cannot take place but because those useless grains take up some of the precious space on the stigma that should be occupied by the correct pollen, evolution in many of the more advanced plant families has led away from promiscuity and towards faithfulness. An extreme example of such faithfulness is, as we have seen, that of the Bee Orchid, which by looking and smelling like a female bee attracts only male bees and therefore does not have its time wasted by unsuitable and unfaithful insects of various sorts. The many other forms of deception practised by flowers to attract visitors, of which we have examined a few examples, are all designed for the same purpose: to establish a special relationship and to try to prevent the visitor from straying outside that relationship and seeking satisfaction from other kinds of flower.

Deception is only one of the ways in which flowers try to make pollinators faithful to them. A more important evolutionary development has been the structural modification of flowers into ever more precise mechanisms which can only be worked by specialized visitors. The faithful relationships between different species of bees and plants have been studied in the greatest detail because of their economic importance, but many types of flower have developed to suit the tastes, needs and talents of completely different kinds of creatures. Botanists recognize separate and distinct types of blossoms designed for particular groups of visitors and characterized by special shapes, devices, colors and smells. There are beetle flowers, fly flowers, honeybee flowers, bumblebee flowers, butterfly flowers and moth flowers, all with their own special features intended to please and satisfy their particular customers and stop them from looking elsewhere.

There are also blossoms which are visited by vertebrates such as birds and bats, particularly in the tropics. Because most of the earlier studies of pollination were carried out in temperate climates, where such visits rarely if ever occur, relationships of this kind were neglected, or even disbelieved, for a long time, and many discoveries in this fascinating field have only recently been made. Bird-flowers are described as being usually vivid in color, particularly red, because birds have much the same color vision as human beings; they often hang down and have no landing platform, because most of the birds concerned, such as hummingbirds, hover in the air; in some cases, for instance the Bird-of-Paradise Flower *(Strelitzia reginae)*, a special perch is provided for birds that need a foothold; the wall is tough, to withstand hard beaks; there is no scent, since birds have scarcely any sense of smell; nectar is very abundant, because birds need a great deal, and is often produced in long spurs to suit their long bills and tongues; and there is no need for a nectar guide, since birds are intelligent enough to find the entrance to the flower without help. Even with such elaborate methods used by those flowers to lure birds and keep them faithful, however, some birds allow their attention to wander. It is reported that in South America a large moth *(Castnia eudesmia)* constantly has to chase away birds from its food plant *Puya alpestris*. However, there are even more examples of a hummingbird being seen driving away hawk moths from its favorite flowers.

It seems to be an inescapable fact that once a creature has formed a close and faithful relationship with one particular type of flower that faithfulness will soon turn into possessiveness. As in all sexual relationships, possessiveness leads to jealousy. Rivals will not only be chased away but if they persist in their attentions may be killed. In any fight between a bird and a moth over the right to possess a flower the odds are heavily in favor of the bird. Its long bill, evolved to enable it to suck nectar from the depths of the blossom, is quite sharp enough to pierce the soft body of a moth; in the duel between the two rivals one has a rapier while the other is unarmed.

In fact it is thought that pollination by birds is a fairly recent development, which originated through more or less accidental visits to flowers in search of prey. Many birds are perfectly capable of catching insects on the wing, but an insect on a flower makes a sitting target. Perhaps long ago a bird taking the lazy way of helping itself to a stationary insect sucking nectar from a flower discovered that the nectar tasted rather good and soon became addicted. Those birds with the longest bills would have an advantage in getting at the nectar in deep flowers and so over long periods of time the typical long-billed nectar-eating birds would have evolved, many of which still eat insects as part of a mixed diet.

Other pollinating birds may once have been fruit eaters, who became impatient and instead of waiting for the fruit to ripen started eating the flowers to get at the sweet nectar inside. From the point of view of the plants this was wholly destructive behavior, which if persisted in would have led to a severe reduction in their breeding potential. It is suggested, however, that some exceptionally skillful and gentle birds with narrow beaks of just the right curvature were able to get at the nectar without doing much damage to the flowers, which were therefore left sexually intact and able to be fertilized. So the plant flourished and multiplied, and the relationship between that particular species of plant and that particular type of bird grew closer and more intimate, each benefiting the other till they became totally dependent on their association, the bird for food and the plant for fertilization.

Some plants avoid damage to their flowers by offering birds

something else to peck, close enough to the blossoms to ensure that pollination is carried out during the process. One such plant is *Freycinetia funicularis*, a member of the Screwpine family *(Pandanacae)* whose fleshy bracts are eagerly eaten by birds of the genus *Pycnonotus*, commonly called bulbuls, which in so doing fertilize the brilliant red, scentless flowers. There is even a type of tree *(Boerlagiodendron)* which is reported to attract pigeons by means of imitation fruit – actually sterile flowers – in among the fertile blossoms; the pigeons can get little satisfaction from the affair, but the flowers get pollinated.

In other cases the association between bird and flower is thought to have originated when birds in hot climates, seeking for something to quench their thirst, drank water which had collected in leaves and blossoms and found the liquid in the blossoms more to their liking because it contained some dissolved nectar. Searching for the source of the sweetness, they probed more deeply into the blossoms till they found the supply of nectar. Birds have a fairly efficient system of communication and soon the discoverers of the new taste thrill would have passed on the information to their kind. Sometimes the relationship between birds and the flowers they pollinate is not a direct one but follows a complex chain in which other creatures are involved. One of the most remarkable stories was told by L. Marden in 1963 in the *National Geographic Magazine*. Under the title "The man who talks to humming birds," he describes how flies are attracted to the flowers of the orchid *Stanhopea graveolens* by the strong odor. The flies are preyed upon by a spider which lies in wait for them. The spider in its turn is preyed upon by a hummingbird, *Glaucis hirsuta*, which in attacking the spider pollinates the flower.

Many other vertebrates have been observed to visit flowers in order to eat them or to suck the nectar: squirrels, tree shrews, bush babies and even a small relative of the kangaroo, *Tarsipes spencerae*, commonly known as the "honey mouse," which lives in West Australia, where its projecting snout and long tongue enable it to get nectar from narrow tubular flowers. There is even a rat *(Rattus hawaiensis)* in Hawaii which regularly pollinates flowers of the tree *Freycinetia arborea* during its nocturnal visits to eat the succulent bracts.

Bat pollinated flowers have their own peculiar characteristics. Because bats lead a nocturnal life, the flowers only come out at night. Since bats are colorblind, the flowers are sometimes drab-colored and often white or creamy so as to be visible in dim light. As bats have a stale kind of body odor, given off by glands and attractive to them socially and sexually, the typical bat-flower gives off at night a strong smell described by some as like that of stale fermentation and by others as sweaty. Because the animals are extremely greedy, the flowers produce great quantities of nectar and of pollen, often from a very large number of anthers. Since bats are big and heavy the flowers have to be wide-mouthed and strong. To enable the animals to cling with their claws, the flowers often grow directly out of main branches, which provide good support (a device called *cauliflory*). Finally, since the bats' echo-sounding equipment does not work very well among foliage, which interferes with it, bat-flowers often hang down on long, pendulous stalks well clear of the leaves (an arrangement known as *flagelliflory*). The peculiarities of bat-flowers are such that few other creatures are either attracted to them or able to visit them, so faithfulness is more or less guaranteed.

We will finish with perhaps the most amazing case yet discovered of total and unbreakable faithfulness between a pollinating animal and a plant. The plant is the fig, and the way it achieves sexual fulfilment is so extraordinary that it would be dismissed as pure fantasy if it appeared in the letter columns of a sex magazine. The mechanism of pollination is so complex and unlikely that the details have only been worked out in the last few years by means of painstaking research, and even now certain points remain obscure.

The juicy "fruit" that we call a fig is in reality a very much modified inflorescence inside a fleshy, hollowed-out receptacle, called a *syconium*. Lining the surface of the cavity are tiny flowers. These flowers are of three kinds: male, female and neuter. The male flowers are reduced to three stamens and little else; the female flowers consist of an ovary topped with a long style; the neuter ones are similar to the female but have a short style. The only entrance to the fig is a tiny orifice at the end opposite the stalk, only just big enough to allow the pollinating insect to squeeze through. This insect is a minute gall wasp, the female of which makes her way in through the narrow hole and lays her eggs in the neuter flowers by sticking her ovipositor through the short style into the ovary at the base. Her ovipositor is not long enough to go down through the long styles of the female flowers; if she tried to, the eggs would not reach the ovary. After she has laid her eggs the female wasp is of no further use and dies. In due course the eggs hatch out in the neuter ovaries. The male wasps, which are wingless, hatch first; they crawl around for a bit and then make for the ovaries containing the young females, which they fertilize before they can get out. After a brief and sexually strenuous life the male wasps die too. The pregnant females emerge from their ovaries, which have become galls, and squeeze out through the tiny orifice, becoming smothered as they do so with pollen from the now ripe male flowers just inside the exit. They then squeeze their way into the next fig, often tearing off their wings in the process, and there repeat their mother's egg-laying procedure, crawling over the female flowers as they do so and spreading pollen on them. After laying their eggs the female wasps die in the dark inside the fig, having only tasted the open air for a brief time between emerging as a pregnant infant from one synconium and entering another as an egg-laying mother to be.

That is a highly simplified version of the sex life of the fig. There are many different species and varieties, each with its own special gall wasp to bring about its mating and each with its own variation of the basic routine. It is reported that some female wasps, instead of merely getting covered with pollen, collect it and put it into special pockets in their bodies before leaving the fig of their birth and entering the one of their death. In some cases, it is reported that other tiny insects get into the figs, some to prey on the wasps and others to lay their eggs in neutral flowers and produce only galls without carrying out pollination. In such cases it is said that the female wasp will deliberately bite off some stigmas to prevent the parasitic visitors from laying eggs that would develop into pests and rivals. Yet again there are varieties of fig, in places where the gall wasp does not exist, which will produce delicious "fruit" without any sexual process at all.

Much remains to be found out about the fig. Meanwhile, because of its construction and its tiny, secret orifice, it is taken by some to represent, according to the doctrine of signatures, the essence of female sexuality. As D. H. Lawrence wrote in his poem *Figs*:

Every fruit has its secret.

The fig is a very secretive fruit.
As you see it standing growing, you feel at once it is symbolic:
And it seems male.
But when you come to know it better, you agree with the
  Romans, it is female.

The Italians vulgarly say, it stands for the female part; the
  fig-fruit:
The fissure, the yoni,
The wonderful moist conductivity towards the center.

Involved,
Inturned,
The flowering all inward and womb-fibrilled;
And but one orifice.

Index

Index